Asset Management Planning and Reporting Options for Water Utilities

Awwa Research Foundation

About the Awwa Research Foundation

The Awwa Research Foundation (AwwaRF) is a member-supported, international, nonprofit organization that sponsors research to enable water utilities, public health agencies, and other professionals to provide safe and affordable drinking water to consumers.

The Foundation's mission is to advance the science of water to improve the quality of life. To achieve this mission, the Foundation sponsors studies on all aspects of drinking water, including supply and resources, treatment, monitoring and analysis, distribution, management, and health effects. Funding for research is provided primarily by subscription payments from approximately 1,000 utilities, consulting firms, and manufacturers in North America and abroad. Additional funding comes from collaborative partnerships with other national and international organizations, allowing for resources to be leveraged, expertise to be shared, and broad-based knowledge to be developed and disseminated. Government funding serves as a third source of research dollars.

From its headquarters in Denver, Colorado, the Foundation's staff directs and supports the efforts of more than 800 volunteers who serve on the board of trustees and various committees. These volunteers represent many facets of the water industry, and contribute their expertise to select and monitor research studies that benefit the entire drinking water community.

The results of research are disseminated through a number of channels, including reports, the Web site, conferences, and periodicals.

For subscribers, the Foundation serves as a cooperative program in which water suppliers unite to pool their resources. By applying Foundation research findings, these water suppliers can save substantial costs and stay on the leading edge of drinking water science and technology. Since its inception, AwwaRF has supplied the water community with more than $300 million in applied research.

More information about the Foundation and how to become a subscriber is available on the Web at **www.awwarf.org.**

Asset Management Planning and Reporting Options for Water Utilities

Prepared by:
Mike Matichich, **Ron Booth**, **John Rogers**, **Eric Rothstein**, **Elisa Speranza**, **Cody Stanger**, and **Ed Wagner**
CH2M HILL
13921 Park Center Road, Suite 600, Herndon, VA 20171
and
Paul Gruenwald
American Appraisal Associates
411 East Wisconsin Avenue, Suite 1900, Milwaukee, WI 53201

Sponsored by:
Awwa Research Foundation
6666 West Quincy Avenue, Denver, CO 80235-3098

Published by:

Awwa Research Foundation

DISCLAIMER

CONTENTS

TABLES

FIGURES

FOREWORD

The Awwa Research Foundation is a nonprofit corporation that is dedicated to the implementation of a research effort to help utilities respond to regulatory requirements and high-priority concerns of the industry. The research agenda is developed through a process of consultation with subscribers and drinking water professionals. Under the umbrella of a Strategic Research Plan, the Research Advisory Council prioritizes the suggested projects based upon current and future needs, applicability, and past work; the recommendations are forwarded to the Board of Trustees for final selection. The foundation also sponsors research projects using an unsolicited proposal process through its Collaborative Research, Research Applications, and Tailored Collaboration programs; and various joint research efforts with organizations such as the U.S. Environmental Protection Agency, the U.S. Bureau of Reclamation, and the Association of California Water Agencies.

This publication is a result of one of these sponsored studies, and it is hoped that its findings will be applied in communities throughout the world. The following report serves not only as a means of communicating the results of the water industry's centralized research program but also as a tool to enlist the further support of the nonmember utilities and individuals.

Projects are managed closely from their inception to the final report by the foundation's staff and large cadre of volunteers who contribute their time and expertise. The foundation serves a planning and management function, and it awards contracts to other institutions such as water utilities, universities, and engineering firms. The funding for this research effort comes primarily from the Subscription Program, through which water utilities subscribe to the research program and make an annual payment proportionate to the volume of water they deliver. Consultants and manufacturers subscribe based on their annual billings. The program offers a cost-effective and fair method for funding research in the public interest.

A broad spectrum of water supply issues is addressed by the foundation's research agenda: resources, treatment and operations, distribution and storage, water quality and analysis, toxicology, economics, and management. The ultimate purpose of the coordinated effort is to assist water suppliers in providing economically and reliably the highest quality of water possible. The true benefits are realized when the results are implemented at the utility level. The foundation's trustees are pleased to offer this publication as a contribution toward that end.

Walter J. Bishop
Chair, Board of Trustees
Awwa Research Foundation

Robert C. Renner, P.E.
Executive Director
Awwa Research Foundation

ACKNOWLEDGMENTS

The research team is very grateful for the time, energy, insights, and evaluations contributed by the participating utilities. The 11 utilities identified below worked as true partners during the conduct of this project. In addition to contributing data for a sample of the assets of their systems, these utilities worked as part of the project team to shape the options being studied, identify and refine the performance measures for study, and identify the evaluation parameters used to rate the options. The individuals identified below served as utility representatives at the focus meetings held for the project and led their utilities' participation in the project. In most cases, these representatives were supported by teams of engineers, planners, financial planners, and utility managers, who helped to develop the information required for the project test, and who participated in other activities such as selecting assets for study, development of data hierarchies, and conduct of project evaluations. Focus group representatives included:

City of Ann Arbor, Michigan, Water Utilities:
Craig Hupy, Systems Planning Manager;
City and County of Honolulu Board of Water Supply:
Clifford S. Jamile, Manager and Chief Engineer;
Ottawa (Ontario) Transportation, Utilities and Public Works Department:
Devesh Shah, Program Manager (Maintenance Management);
Pittsburgh Water and Sewer Authority:
Kelly Sikorski, Finance Director;
City of Portland (Oregon), Bureau of Water Works:
David Evonuk, Project Manager, Distribution System Master Plan;
Portsmouth, Virginia, Department of Public Utilities:
James Spacek, Director
Saint Paul Regional Water Services:
Christine Meyer, Information Systems Manager
San Francisco Public Utilities Commission:
William H Laws, Rate Administrator;
Santa Clara Valley Water District:
Alan Zeisbrich, Senior Project Manager;
Seattle Public Utilities:
Elizabeth S. Kelly, PE, Strategic Operations Division Director;
Toronto (Ontario) Water and Wastewater Services:
Michael D'Andrea, Manager, Infrastructure Asset Management.

In addition, the research team appreciates the insights, guidance, and comments provided by AwwaRF's staff and the Project Advisory Committee assigned to this project. AwwaRF's efforts on this project were led by project manager Robert D. Allen. The Project Advisory Committee members, who attended some project workshops, and also provided useful and timely comments on project progress reports and other points of review, are identified below:

John Dyksen, East Coast Water Practice Leader, Black & Veatch
Craig W. Charleston, Assistant Administrator, Columbus, Ohio, Division of Water
David Muto, Manager of Water Supply and Treatment, Seattle Public Utilities

The project team was fortunate to have as a research advisor Gordon Logan, Manager of Water Asset Planning, at South East Water, Ltd. South East Water, an Australian utility, is a leader in the development and implementation of performance-based asset management. Gordon attended all three of the project workshops and provided useful presentations and review comments on the parameters and sample output formats developed for the study. Mr. Logan also contributed several valuable examples for inclusion in the appendix.

The team also appreciates the insights from the vantage point of the financial community that were provided by James Wiemken, Director for Public Finance at Standard and Poor's. Mr. Wiemken made a valuable presentation at a project workshop that is included as part of the appendix to this report, along with a related white paper on the market implications of utility financial planning that was developed for a recent AWWA study. The white paper is also included in the appendix; thanks to AWWA for permission to include it.

EXECUTIVE SUMMARY

PURPOSE

This project was designed by AwwaRF as a management study to provide comparative case studies in asset management reporting. Case studies were developed for 11 utilities to illustrate the outputs that could result from adopting varying levels of asset management programs for a sample of their assets. The study compares side-by-side the reports that would be generated by basic and higher-level asset management programs. The appendix includes sample outputs for the three asset management options studied for the 11 participating utilities and evaluations of the options and outputs by the utilities.

This report is intended to aid utilities in developing responsive asset management strategies that comprehensively address utility infrastructure planning, with an emphasis on infrastructure renewal. The examples and analyses in this report are intended to help water utilities identify asset management options that are worth considering and performance measures worth tracking. Information is also provided on data development and maintenance costs for the reporting options, and potential benefits of the information for management decision-making and advocacy for appropriate renewal and replacement funding with governing bodies and other stakeholders.

ASSET MANAGEMENT OPTIONS STUDIED

Working closely with 11 participating utilities, the research team for this project defined and then refined three levels of asset management planning for detailed evaluation:

- **Basic Option**, in which traditional concepts of valuation and service life are used to approximate condition, estimate renewal and replacement factors, and help target renewal expenditures. Key concepts used in this option are the replacement value of the assets and realistic estimates of the useful lives of the assets.
- **High-End Option**, in which key performance measures, such as service outages and regulatory compliance, are identified and prioritized. Performance scores are then used to assess the condition of individual and grouped assets to help guide renewal and replacement decision-making.
- **Strategic Option**, which incorporates elements of the basic and high-end options but also uses additional features to estimate the financial impacts of renewal and replacement decision-making and its effect on system value and condition over time.

METHODS USED IN THE STUDY

To develop and then refine the three levels of asset management planning that were studied, the project research team and participating utilities worked collaboratively, including selecting performance measures for study and developing appropriate hierarchy structures. We also collaborated on the development of an evaluation framework to assess the value of the options studied. As such, the participating utilities worked as true partners in the execution of this project. The project team selected seven performance measures out of nearly 50 candidate options to test for the high end and strategic options. Based on the specific options selected for study, the research team developed a spreadsheet template model that was used to produce sample tables and graphic outputs for up to 1,000 asset records contributed by each of the participating utilities. Scorecards and letters of comment were used to evaluate the usefulness of the options and outputs in terms of value to utility staffs in formulating recommended asset management programs and of making the case to governing boards and other stakeholders for required resources to support effective asset management.

The study presented in the report is not a software development project; the template model used to generate the sample outputs was developed as a testing platform to illustrate the types of outputs that would be generated by data samples of up to 1,000 records. Readers of the report might choose a variety of software approaches to implement reporting options that they find valuable among the options presented and evaluated. For example, the participating utilities employed a set of diverse software implementation approaches to incorporate lessons learned from this project, including:

- Incorporating reporting capabilities from the samples shown in this report into existing systems already populated or partially populated by the utility, such as geographic information systems (GIS) or computerized maintenance management systems (CMMS).
- Purchasing and implementing freestanding asset management software that emulates the reporting outputs found to be valuable, or which has the capability to be adapted to include reporting capabilities found to be valuable.
- Developing customized systems that emulate the reporting capabilities of outputs found to be of high value in database or other programming environments.

FINDINGS AND RECOMMENDATIONS

Key overall findings and recommendations from the project are summarized below.

Options, Performance Measures, and Outputs Were Found To Be Valuable Decision Aids

All three of the asset management options provide substantial value above no action or current practices for both plant and pipe assets. On a scale of 1 to 5, with 5 being a perfect score, participating utilities gave all options a rating of at least 3.3 (Figure ES.1).

In general, participating utilities found the use of prioritized performance measures to be critically important in creating an effective asset management program. Other useful features included the ability to set system-specific weights on the performance measures, and to customize the scales used to "score" asset performance. As detailed in Chapter 9 (Findings), the study found some of the seven performance measures to be of substantial value. Several measures received ratings of 4.0 on a 5.0-point scale from the participating utilities either in terms of their value in aiding management decision-making or in making the case for required funds to governing boards and other stakeholders.

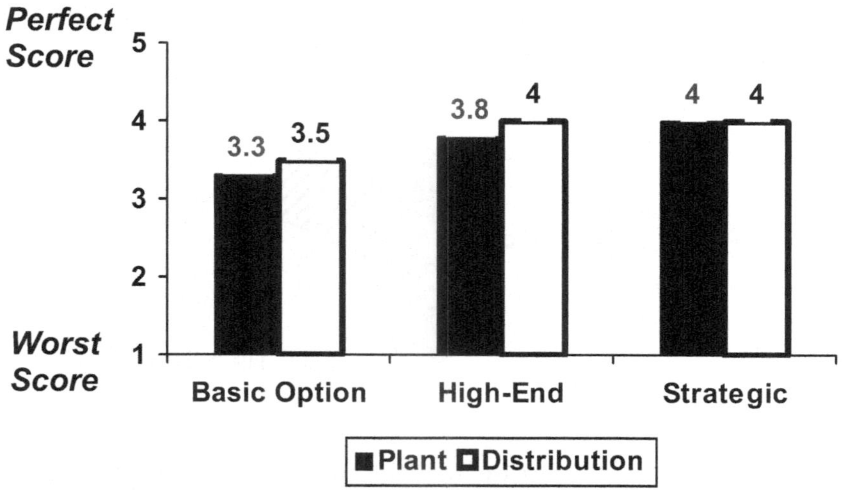

Figure ES.1 Overall score of plant and pipe assets by option

Data management challenges were considered much greater for the high-end and strategic options (Figure ES.2). Many of the participating utilities found that the input information required to populate the samples for the basic option was generally available, although in many cases requiring work to integrate information across data systems. The performance information to populate the performance measures for the high-end and strategic options in many cases required much more significant data generation and management efforts. Many of the participating utilities thought that organizational or cultural changes would be required to successfully implement these options throughout their systems.

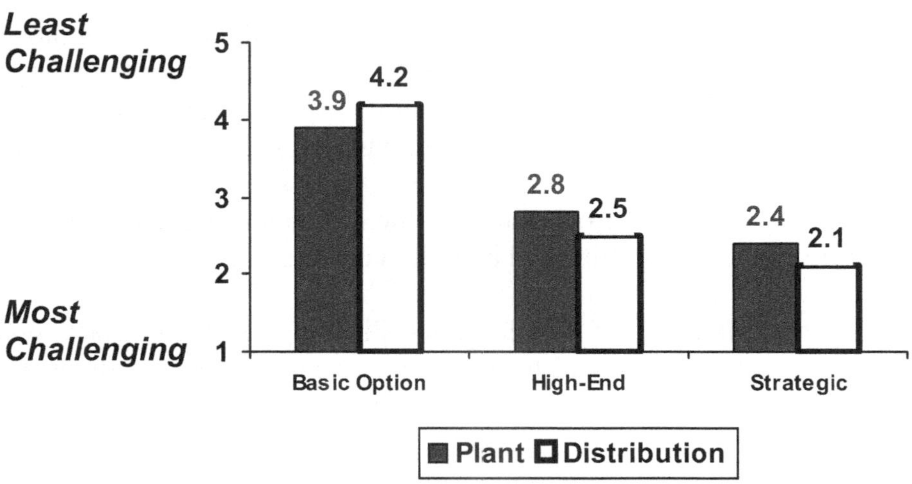

Figure ES.2 Level of data management challenge for plant and pipe assets by option

Despite the additional costs and staff resources associated with implementing the high-end and strategic options, most of the participating utilities have indicated that they intend to implement asset management strategies that emulate either the high end or the strategic options studied in this project (Figure ES.3). They deemed the information and outputs for these options to be worth the additional draws on staff and other resources.

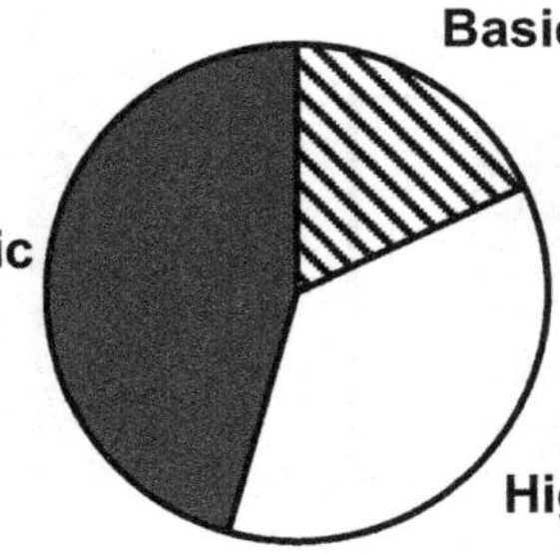

Figure ES.3 Proportion of participating utilities planning to implement basic, high-end, and strategic programs

The diverse perspectives of the participating utilities illustrate the context-specific nature of developing an appropriate asset management strategy. There are some contexts in which the basic option represents an appropriate strategy. Factors such as the availability and complexity of system data and the number of stakeholder groups and their specific information needs strongly influence whether it is appropriate to undertake the additional data development and maintenance required to support the high end and strategic options. For example:

- Clifford Jamile, Manager and Chief Engineer for the Honolulu Board of Water Supply, concluded that "Since we have implemented most critical-need projects identified through our previous asset management efforts, we find substantial value in the capabilities of the strategic option to help identify the next tier of renewal and replacement projects for attention. As such, in our case it is worth the additional data development costs to have the benefit of the outputs of the strategic option."
- James Spacek, Director of the Portsmouth, Virginia, Department of Public Utilities, concluded that "The City of Portsmouth has adopted the 'basic option' approach for its utility systems.... Both the 'high-end' and 'strategic' options have compelling potential benefits, but for our situation, the cost and complexity seem to represent an effort with somewhat diminished benefit in our case."

Gordon Logan, Manager of Water Asset Planning at South East Water, LLC, in Australia, who served as special advisor to this project, helped to put the diverse choices of the participating utilities into perspective with the following comments:

> The asset management reporting options under consideration for this project can be seen as a natural progression in an asset management continuous-improvement strategy. The initial option adopted by an individual utility will be dictated by the amount of asset-related data available at this time. An asset management–strategic plan will then consider which option the utility will move towards and over what timeframe. Regardless of which initial option is selected, in order to develop an acceptable asset management strategy, there should be sufficient information to determine a picture of requirements over the entire asset portfolio.

As detailed in the scorecard ratings and letters of comment included in the study and summarized in Chapter 9 (Findings), several of the sample graphics and tables developed for this project were found to be valuable in developing staff recommendations and in making the case to governing boards and stakeholders. Just as the overall direction forward varied significantly among the participating utilities, the usefulness of the outputs was found to vary, too. The commentary on the performance measures and outputs by the participating utilities provides useful insights that can help readers to consider the usefulness of the sample outputs for their situations.

Data Management Challenges Merit Close Attention

Because the data-gathering and development challenges associated with populating the 1,000-record samples were even more challenging than had been anticipated, many of the recommendations for future direction relate in one way or another to data management efforts. Many of these recommendations center on the theme that water utilities need to be selective about how they implement asset management programs. Examples of strategies that some of the participating utilities intend to employ when moving forward with implementation of their asset management programs are the following:

- Implement the selected asset management option in phases. There are several variations of this strategy. In some instances, this will be done by addressing first assets of a certain type (e.g., all pumps or large water mains). In other cases, by addressing all assets within a limited geography or by addressing those assets deemed most important through a preceding criticality analysis.
- Start by implementing new protocols for capturing desired information assets being added to or renewed or replaced within the systems
- Carefully investigate opportunities to mine existing databases, such as GIS and CMMS systems, either to capture information that would be imported into asset management systems or by building additional reporting capabilities directly into these systems.
- Strategically justify the addition of each performance measure and other attributes maintained in the asset management databases. Do not continue to gather data for legacy measures unless they are found to continue to provide value.
- Start at a high level of hierarchy, with summary-level performance and financial information, and then building out detailed information by geography or other criteria as staff time and other resource availabilities allow. Critically consider the benefits and associated costs of capturing additional levels of detail.
- Consider conducting a beta test for a small sample of assets, such as the ones used for this study. Such samples can provide valuable insights that can be used to optimize data gathering and process efforts during system-wide implementation. Conducting such beta tests also increases the likelihood that selected output measures will be useful.

The appropriate data platform to store and analyze the information was also found to be system-specific, depending on such factors as the availability of data systems and whether data protocols exist within utility operations or across all departments within a City, County or other government entity. As detailed in the letters of comment included in the Appendix CD-ROM, data management strategies for moving toward system-wide asset management implementation range from developing customized systems, to purchasing and implementing dedicated asset management software, to building additional customized capabilities into existing maintenance management systems or geographic information systems.

Developing an Appropriate System-Specific Hierarchy is Key to Success

There are several ways the assets can be grouped within the asset management databases (for example, by location, by type of asset, by functioning system). The appropriate method is system-specific, depending on such factors as how assets are maintained and how capital projects are organized for inclusion in CIP and budget documents. The selection of a hierarchy had a significant influence on the usefulness of some of the reports and figures tested for decision-making within participating utilities. As detailed in the letters of comment included in the Appendix CD-ROM, several of the participating utilities made adjustments in their approach to hierarchy development, based on their review of outputs that resulted from the project tests.

Stakeholder Profile and Decision-Making Culture Influence Both High-Level Strategies and Detailed Program Implementation

The participating utilities identified a broad spectrum of stakeholder groups with an interest in asset management decisions, including governing boards, municipal governments, the financial community, the media, customers, and employees. The identified stakeholders varied not only by size of the system, but also by regulatory context and other factors. For example, in some cases, state public utilities commissions have a strong role through their oversight and approval role related to budgets and rate increases. In other cases, state agencies have no or only an advisory role on such matters. Utilities that finance substantial portions of their capital programs through municipal bonds identified bond-rating agencies as key stakeholders of interest. The participating utilities found it to be valuable to think of specific stakeholder groups, and even the most influential individuals within these groups, when considering which tables, graphics, or other communication media will be most valuable to successfully make the case for resources needed to implement appropriate asset management strategies.

Ancillary Benefits to Action on Asset Management Can Be Substantial

In addition to helping utility staff and management to identify appropriate levels of renewal and replacement investment, developing asset management programs that emulate the options tested were found to have substantial additional value for water utilities. As discussed in an appended letter report from American Appraisal Associates (AAA), the data developed to support these options may help align fixed asset registers to more accurately reflect the actual assets in service and their value. Paul Gruenwald, Vice President of AAA and Managing Principal for the firm's Milwaukee Property Appraisal Services Group points out in the letter report: "From experience in appraising water utilities, we find that what appears in the capital asset ledger maintained by the Finance Department is different than what is actually in the plant. This is because there are normally no established procedures for plant personnel who are familiar with what is occurring in the plant to report retirements and additions to the Finance Department.

If an electronic interface could be established that would allow plant personnel to automatically report to the Finance Department assets that are being constructed, replaced, or disposed of, it would improve the accuracy of the capital asset ledger, reduce the manual effort in the Finance Department to capture this information, and allow the Finance Department to report better information."

James Wiemken, Standard & Poors sector leader for municipal water and sewer utility ratings and state revolving fund ratings throughout the United States, notes that developing capital and rate programs that provide for long-term sustainability of a utility system's assets is viewed favorably by credit rating analysts. In his appended white paper, Mr. Wiemken points out: "True rate stability and transparency assumes that a system's current and likely future needs have been measured and are relatively known. The average increase in rates to be targeted over the next decade cannot be known without some idea of the cost pressures a utility may face, and without an honest effort to estimate these needs, it will be extremely difficult to educate and inform ratepayers."

Detailed Findings and Recommendations Provide Additional Guidance

The body of this report provides many additional findings (Chapter 9) and recommendations (Chapter 10) related to the options and strategies evaluated in this study. In addition, more detailed discussions of some technical study topics include:

- Selection and use of appropriate performance measures (Chapter 4)
- Data development and management (Chapter 5)
- Developing appropriate data hierarchies (Chapter 6)
- Determining an appropriate evaluation framework (Chapter 7)
- Defining the relationship of asset management to related utility planning processes (Chapter 8)

In addition, the appendix contains sample outputs for the three asset management options, and scorecard ratings and letters of comment contributed by the participating utilities, creating a case book of examples. The utilities' examples and lessons learned can be used to gain additional insight into how and when the options studied in this project will prove most valuable.

CHAPTER 1
INTRODUCTION

CHALLENGES TO AND DRIVERS FOR CONTINUED WATER INDUSTRY ACTION ON ASSET MANAGEMENT

As water utilities strive to implement appropriate strategies for managing their considerable treatment plant and distribution assets, they are faced with challenges, two of which are fully valuing water in a marketplace whose customers and policymakers have become accustomed to below market rates for service and developing more strategic asset management programs in the face of increased funding competition.

As detailed in such recent studies as AWWA's *Dawn of the Replacement Era* and EPA's gap analysis studies, it is increasingly understood that historic under-funding of renewal and replacement of water and wastewater systems has led to a decline in the overall condition of assets—now reaching a critical stage. Critical system components are requiring costly maintenance and failing outright with increased frequency. Because of this, water is undervalued in the market place. As detailed in the recent AWWA study *Avoiding Rate Shock: Making the Case for Water Rates*, the fact that customers have traditionally not paid the full life-cycle costs for water increases the challenge in getting support for all of the funds that are needed to maintain water system assets in sound condition.

Utility managers interested in developing more strategic and proactive asset management programs are often stymied by competition for funding. Key examples are the following needs:

- To address added security requirements being demanded by stakeholders and regulators since the September 2001 terrorist attacks
- For some systems to devote substantial resources to compliance-mandated upgrades to treatment processes (e.g., to address requirements of the 1996 Safe Drinking Water Act Amendments)
- For capacity expansion resulting from growing service populations

Fortunately, important countervailing drivers are at work to move utility asset management from its historically reactive mode to proactive development and implementation of cost-beneficial strategies. These important drivers for action include the following four:

Water Industry Research

Water industry studies, such as the Research Foundation's Nessie Curve studies, the EPA Gap studies, and the more recent General Accounting Office (GAO) studies, objectively document the magnitude of the asset management challenge. While these multijurisdictional studies do not identify system-specific remedial strategies, they effectively call attention to the substantial needs for increased renewal and replacement spending. These highly visible, scene-setting efforts lay a strong foundation for industrywide leadership to ratchet up the pace of replacement.

Guidance documents and data specific to the water-wastewater industry have recently been developed that provide an important framework for water industry action. Chief among these include:

- The *Capital Planning Strategy Manual* (Beaudet *et al.* 2001), which provides an important framework for evaluating and prioritizing renewal and replacement projects in relation to other competing capital projects (such as upgrade and capacity expansion projects). The ability to prioritize objectives and develop a ranked listing of projects in relation to their contribution to a utility's key objectives provides a critical framework for weighing the relative merits of competing renewal and replacement projects.
- *Managing Public Infrastructure Assets to Minimize Cost and Maximize Performance* (AMSA 2002) provides guidance in the development of asset hierarchies, data management strategies, and other aspects of formulating an asset management program. The handbook was rolled out in a series of workshops throughout the United States.
- Research and application of infrastructure asset management tools and approaches in other countries, such as England and Australia, which provide useful inputs to the refinement of approaches in North America.
- *Selection and Definition of Performance Indicators for Water and Wastewater Utilities* (Crotty 2004), which identifies recommended categories of performance measures and suggested indicators within those categories to achieve a balanced overall performance measurement program.

Other-Industry Research

Research performed in other industries in establishing asset management approaches and strategies provides useful guidance that can be adapted by the water industry. Key among this work is the development of an asset management reporting option for general infrastructure assets such as roads and bridges by the Government Accounting Standards Board (GASB) in Statement 34, issued in June 1999. Many water utilities are organized as enterprise funds and have elected not to implement the asset management reporting option. However, regardless of what method utilities use to satisfy formal financial reporting requirements, every utility has the opportunity to enrich its asset management strategy by considering the product of GASB's 10 years of deliberations—the asset management option defined in Statement 34.

Data Management Advances

Recent technological advances enable more efficient automated and integrated data management and analysis than ever before. Powerful microcomputer database programs can now be coupled with Web-based data-sharing capabilities and enhanced supporting software applications (e.g., GIS and CMMS). These integrated data capabilities provide the opportunity to efficiently record, store, and analyze information related to utility system assets in a manner that far exceeds what was possible even 5 years ago.

Development of Utility-Specific Asset Management Applications

The water treatment plant inventory and assessment tool developed in AwwaRF Project 460, *Water Treatment Plant Infrastructure Assessment Manager: User's Manual* (Elliott *et al.* 2001) provides the capability to inventory and store information about the current condition of a detailed list of water treatment plant assets. A number of consulting firms have developed asset management software systems that provide the capability to track the condition of utility system assets over time, based on varying assumptions of renewal and replacement investment.

These drivers provide the basis for management empowerment in the asset management arena—the capacity for managers to develop better-informed strategies that maximize cost-effective levels of renewal and replacement investment and, importantly, to make the objective, well-supported case necessary to win support from stakeholders and governing boards.

The potential for utility management empowerment as a result of these drivers has been realized only partially to date. Many utility managers, faced with budget pressures and the need to expand their systems or implement compliance-driven capital projects, have been reluctant to implement some of the emerging approaches for asset management. For example, considerable interest has been expressed in the asset management approach developed by GASB for its asset management reporting option; however, no major utility system in the United States has fully implemented an asset management system that follows the specifics of GASB's identified approach. In addition, while the guidance documents cited above and others provide increasingly sophisticated grounding in the elements of a sound asset management program, there is limited comparative testing to provide real-world insights into the costs to develop data for the options and to assess the usefulness of the information produced by varying levels of asset management planning. Most of the existing literature is either theoretical or based on single case study applications.

SPECIFIC NICHE SERVED BY THIS PROJECT

This project was designed by AwwaRF as a management study to help address the lack of comparative case studies, by developing case examples that illustrate the kinds of outputs that would result by adopting varying levels of asset management programs for a sample of actual assets. The study is intended to provide side-by-side comparisons of the actual reports that would be generated by basic and higher-level asset management programs, including ones that emulate the asset management option identified in GASB's Statement 34, which includes the requirements to

- Inventory the assets
- Create a replicable condition assessment
- Identify a target condition level for the assets
- Identify the funding requirements to maintain the assets at the target condition level
- Identify recent historical expenditures to maintain/renew the assets

The report appendix includes sample outputs for the three asset management options studied for the 11 participating utilities shown in Figure 1.1.

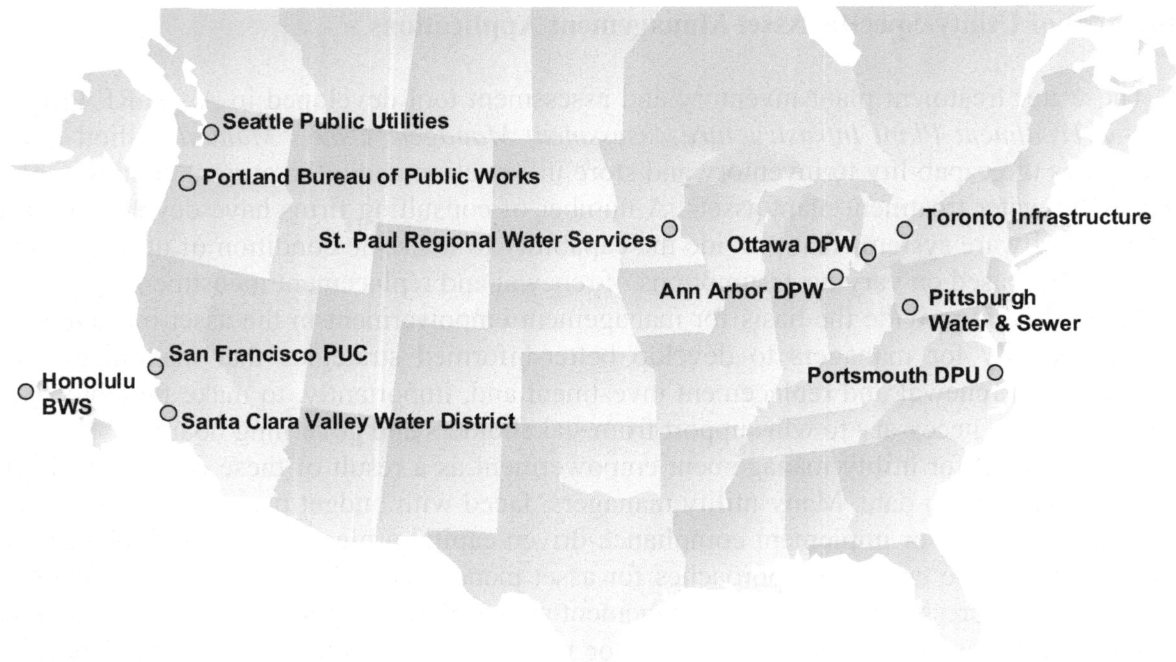

Figure 1.1 Map showing the 11 participating utilities' locations

These 11 geographically diverse and different-sized utilities participated in the development of approaches and methods throughout the project. As illustrated in Figure 1.2, they functioned as true partners in the research effort. In addition to contributing data for a sample of up to 1,000 assets, they helped to define and refine the options and performance measures studied and helped to define and refine the tables and figures used to illustrate the outputs for the options studied. Options for addressing elements of the GASB Statement 34 methodology (e.g., condition assessment, defining the financial impact of achieving a target condition level) were considered and tested, where applicable.

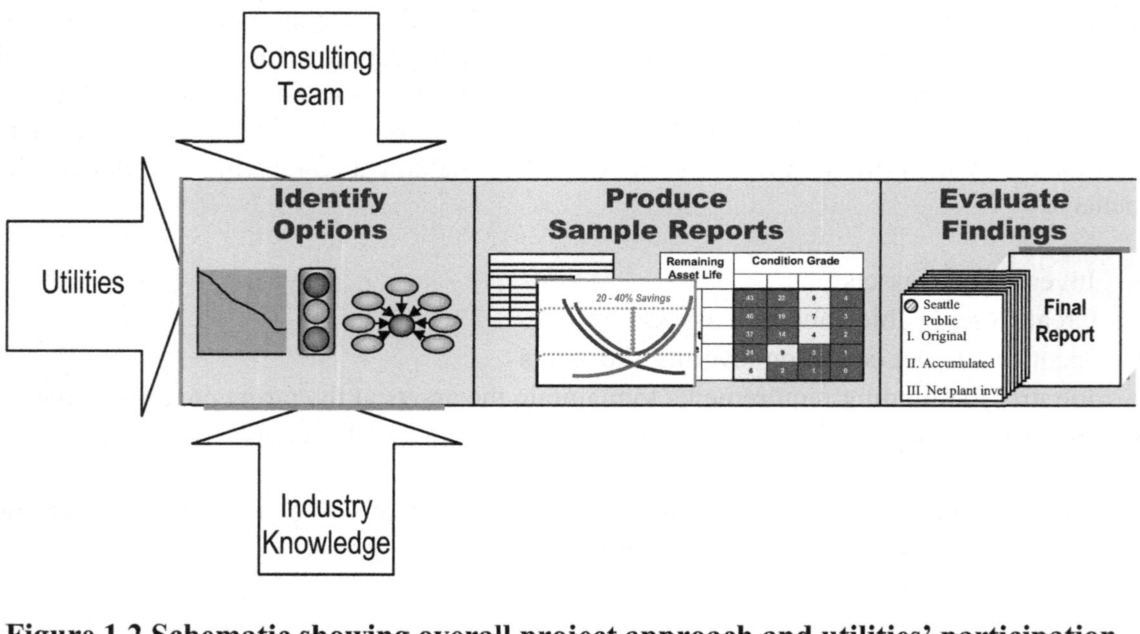

Figure 1.2 Schematic showing overall project approach and utilities' participation

The participating utilities also contributed evaluations of the options and measures studied. These evaluations, which include scorecard evaluations and letters of comment, provide important context for considering the usefulness of the options, measures, and outputs developed and evaluated in this study.

The evaluations by the participating utilities along with the findings and recommendations of the study provide a framework that water utilities can use to identify an appropriate asset management option and implementation strategy, taking into consideration such factors as data and staffing requirements and the usefulness of information and reporting outputs to decision-makers and likely stakeholders specific to the system. The study identifies a number of implementation strategies, such as geographic- or facility-based options, that the utility partners identified as likely follow-up steps in their extending the work done during the course of this project to more widespread application within their systems.

The study presented in the report is not a software development project. The project research team developed a spreadsheet-based template with the ability to process information for a limited sample of up to 1,000 records in order to illustrate the reporting outputs for the options and performance measures developed for study. Readers of the report, like the participating utilities themselves, might choose a variety of software approaches to implement reporting options that they find valuable among the options presented and evaluated in this report. Software implementation approaches could include:

- Incorporating reporting capabilities that are found to be valuable from the samples shown in this report into existing systems already populated or partially populated by the utility, such as GIS or CMMS
- Purchasing and implementing of freestanding asset management software that emulates the reporting outputs found to be valuable, or which has the capability to be adapted to include reporting capabilities found to be valuable
- Developing customized systems that emulated the reporting capabilities of outputs found to be of high value in database or other programming environments

As further detailed in some of the letters of comment prepared by the participating utilities for inclusion in the report, the selection of a specific hardware and software solution is context specific, depending on such factors as the size of the utility and number of assets, other information systems that have already been implemented (e.g., geographic information systems or computerized maintenance management systems), and other utilities or departments with which the utilities may share computerized hardware, software, or data. Some of the utilities have indicated that they intend to build reporting capabilities that they have found valuable through their participation in this study into CMMS or other existing systems; some are in the process of developing customized asset management systems to address their needs; others have indicated that they are likely to purchase and implement asset management systems from outside vendors.

This report is intended as an aid to utilities in developing responsive asset management strategies that comprehensively address utility infrastructure planning, with an emphasis on infrastructure renewal. Reviewing the examples and analyses in this report is intended to help water utilities to identify asset management options worth considering, performance measures worth tracking, data development and maintenance costs for the options, and potential benefits of the information produced as aids to both management decision-making and making the case for appropriate renewal and replacement budgets to governing bodies and other stakeholders.

CHAPTER 2
METHODS USED IN THE STUDY

This section describes the primary methods that were used to conduct this study. Primary elements of the approach, detailed in this chapter, are the following:

- Development of initial statement of asset management options and measures to study
- Refinement of asset management options and measures
- Development of participating utility test data
- Development of template model
- Refinement of parameters and data
- Development of evaluation framework

DEVELOPMENT OF INITIAL STATEMENT OF ASSET MANAGEMENT OPTIONS AND MEASURES TO STUDY

During the initial phase of the project, the principal investigator (PI) identified three asset management options for consideration by the project team, including rough samples of some of the potential output figures and tables for the three options. These preliminary options were based upon the following:

- Prior asset management programs developed by the PI and other members of the project research team
- The literature review conducted for the project
- A preliminary survey of the current and intended state of asset management that was completed by the participating utilities during the first month of the project. The survey addressed such topics as the current state of practice within the utilities, availability of information, drivers to action, and areas deemed most in need of improvement.

As mentioned in previous chapters, the three options presented for consideration are basic, high-end, and strategic:

- **Basic** is based on depreciation and useful life principles and updated to reflect more-accurate price and useful life considerations. Elements of this approach would include inventory of assets, historical investments and unit prices, and useful life schedules. Typical outputs for this approach would include replacement valuation, accumulated depreciation, remaining undepreciated value, and annualized renewal and replacement factors.
- **High-end** is based on performance measures that are used to estimate the condition and relative renewal and replacement needs of system assets. High-end programs were defined to include performance and condition assessments at a single point in time. This option includes inventory and valuation of assets, as in the basic option, but would also include identification and prioritization of key performance criteria, performance measurement factors, and condition assessments. Typical outputs for

this option would include individual asset performance databases and summary measures of performance and condition by asset, by subsystem, and for the system overall.

- **Strategic** is based on using data integration, risk management tools, and performance assessments to provide a further refined analysis. This option includes all the elements of the high-end option plus the following: forecasts of conditions over time in response to renewal expenditures, maintenance practices, and other policy decisions; risk analysis and scenario analysis; and greater integration of data. Typical supplemental outputs for the strategic option would include condition assessment over time and comparative scenario analyses that take risk and probability into consideration.

REFINEMENT OF ASSET MANAGEMENT OPTIONS AND DEVELOPMENT OF SPECIFIC MEASURES TO TEST

The preliminary definition of asset management options developed by the PI was presented to the project team at the first project focus group workshop, which was held in early 2003. The options were refined during that workshop, during monthly project conference calls with the focus group and research team members, and during the second workshop, which was held in September 2003. The refinements were made to reflect input from the focus group of utility participants and the other members of the research team and to provide details required to create options capable of being tested. The final resulting options and outputs are detailed in Chapter 3. Examples of the refinements that were made during the course of the project are:

- Recasting the basic option as a service-life rather than a depreciation-based option to avoid confusion with the original-cost-based depreciation analyses typically prepared as part of financial statements/fixed asset databases.
- Identifying the specific performance measures worth testing for the high-end option. As detailed elsewhere in this chapter, seven measures that were studied by all of the participating utilities were selected from an initial list of more than 30 candidate measures. Several of the participating utilities evaluated additional, supplemental performance measures.
- On the basis of feedback from the participating utilities regarding the availability of input data, the strategic option was recast as an extension of the high-end option to include a projected future timeline and to allow scenario analyses that show the expenditure requirements and effect on system value of alternative renewal and replacement scenarios. Except for the few utilities that populated supplemental performance measures on criticality and consequences, most of the participating utilities indicated that they were not able to populate measures that would evaluate risk as part of the study.

DEVELOPMENT OF PARTICIPATING-UTILITY TEST DATA

Each utility was requested to identify a portion of its system that could be used to test the three asset management options. The research team suggested that several criteria be used to select the portion of the system that would be tested:

- Access to information—It was suggested that the utilities select a portion of the system for which basic inventory, financial, and performance information would likely be readily accessible so that there would be some assets with relatively complete information populated in the project databases. In addition, it was suggested that the utilities select some portions of their systems for which there would likely be data challenges (e.g., where there would likely be gaps in identifying installation dates or cost information). This was suggested so that the participating utilities could get a clear sense of the activities that would need to be undertaken (and the resources that would be required) to fully implement the asset management options, even for "difficult" assets.
- Old vs. new—It was suggested that the utilities consider developing information for some older and some newer portions of their systems. This way, it would be possible to compare and contrast the outputs for portions of the system for which renewal and replacement would be more likely with portions of the system for which it would be less likely.
- Potential immediate value—If there was a portion of the system for which development of a renewal and replacement strategy was imminent, it was suggested that this segment of the system be considered for use in the project so that the utility would get immediate value in its investment of resources to gather and develop information for the project.
- Number of records—It was suggested that each utility identify a sample of up to 1,000 records to include in the analysis.[1] This was done to keep the data-gathering effort manageable, both in terms of the project schedule and the level of in-kind hours that the participating utilities agreed to contribute to the project.[2]
- Number of systems—Initially, the research team suggested that the participating utilities limit their data sets to three main assets, with two levels of hierarchy below this, with up to 100 detailed records at the lowest level of the hierarchy. However, given the diversity—in terms of assets and level of detail—of the systems represented, the utilities requested that an optional fourth level be allowed, to permit greater flexibility in defining the level of detail provided. This was done, and the utilities were then enabled to define any number of assets and subassets, with the recommendation that the total number of records be limited to 1,000.

Based on these criteria, the participating utilities defined a fairly diverse set of assets to include in the project. Some decided to include both plant and distribution system assets, whereas a few of the utilities decided to include only one or the other of the two main systems. Some examples of the initial asset groupings developed for study:

- Portland, Oregon, decided to use the headworks facility plus several quarter-section segments of its distribution system.

[1] One of the utilities, the City of Toronto, ended up contributing information for approximately 2,000 records of assets. This was done to allow inclusion of two full segments of the city's distribution system, each with nearly 1,000 records, plus a high-level evaluation of a portion of the city's water treatment plant assets.

[2] It was anticipated that some of the utilities would find it useful to contribute more than the initially committed 120 hours of in-kind support to the project. In fact, the average contribution for the participating utilities was more than 200 hours.

- The Pittsburgh Water and Sewer Authority decided to develop a database that included certain equipment categories (e.g., valves and meters) distributed throughout the system.
- Ann Arbor, Michigan, included the ozone system at its water treatment plant plus a portion of the North East Area Pressure District in its distribution system.
- The Santa Clara Valley Water District decided to include portions of one of its treatment plants and several portions of its treated water transmission and delivery system.
- Toronto included two segments (one old, one new) of its distribution system plus some high-level plant assets.

Once the participating utilities selected the primary asset groups to include, it was necessary to determine how the four levels of hierarchy allowed for the analysis would be used (i.e., at what level of detail to include information for assets in the system). The process of determining appropriate hierarchies, and examples from several of the utilities' data sets, is illustrated in Chapter 6 of this report.

Over the course of the project, some of the participating utilities revised the assets included in the study and the use of the levels in the hierarchy. The final overall assets included in the study for each of the participating utilities, and the hierarchies employed, can be observed by reviewing Table B-1 in the sample outputs for each utility in the appendix.

During the first project workshop meeting, the research team and participating utilities identified 47 candidate performance measures for consideration. (See Table 2.1 on p. 14 for a sample of the measures considered.) The team then prioritized the measures to identify a subset of seven measures for which data would be gathered for this project. As detailed in Chapter 4, several of the utilities gathered information for several additional performance measures that they deemed particularly important to their consideration of asset performance and the development of renewal and replacement strategies for their systems. Over the course of the project, the commonly studied performance measures were refined in response to further evaluation and analysis by the utilities and the research team. Ultimately, the following performance measures were included in the study for all of the utility participants:

1. Service reliability—interruptions
2. Service reliability—main breaks (per mile of pipe)
3. Water quality
4. R&R status
5. Maintenance activity
6. Preventive maintenance
7. Age vs. service life

DEVELOPMENT OF TEMPLATE MODEL

The project approach called for development of a spreadsheet application, database application, or the use of other software to test the usefulness and practicality of the candidate asset management options developed for the project. After the first project workshop, the research team decided that developing a spreadsheet template model would be the best way to test the asset management options. This decision was made in light of the following:

- The specific performance measures selected for study for the high-end and strategic options.
- Source information identified by the participating utilities for the background, financial, and performance measures. Several of the participating utilities indicated that they could likely port at least some parts of the required data into a spreadsheet input file from their GIS systems, maintenance management systems, fixed asset records, and other electronic data-tracking and storage systems.
- The flexibility that a spreadsheet format would provide for customizing the studies to some specific requirements and interests of the individual participating utilities (e.g., inclusion of some additional utility-specific performance measures, flexibility in determining how many levels of the asset hierarchy to use).
- The capability that a spreadsheet format provided to allow the participating utilities to assign different weights, based on their priorities, to the performance measures being studied, and to adjust the performance scales that were used to assign points to each asset, based on how it performed. For example, for some utilities a single failure to meet standards might cause asset to lose half of the total possible point score of 10 points, and results in a score for that performance measure of 5; some other utilities may be more tolerant of a single incident of failure to meet standards, and might assign a score of 8 for that performance measure for one incidence of missed standards. (Most of the participating utilities took advantage of the opportunity to customize the policy and rating parameters by adjusting the default weightings to the performance measures or by adjusting the one or more of the performance scales used to assign performance points to assets for each of the performance measures. This was done by changing the weighting and scoring factors in the form illustrated for Santa Clara Valley Water District on p. 37.)

Immediately following the first project workshop, the research team developed and distributed to the utilities a data input file, so that the participating utilities could begin gathering the background, financial, and performance information for their asset samples. Monthly conference calls of the participating utilities were used to provide guidance to the utilities as they developed their data sets, and to refine the data collection tool and study parameters in response to continuing feedback from the participating utilities.

The template model was developed, and a first round of output samples was developed and distributed to the participating utilities for review at the second project workshop.

REFINEMENT OF OPTIONS, PERFORMANCE MEASURSES, AND OUTPUTS

The second workshop meeting for the project was used to gather insights on the first round of sample reports that were developed for each of the utilities. Based on insights and observations from the research team, the participating utilities, and AwwaRF's Project Advisory Committee for the project, refinements were made to the definition of the asset management options being studied, the specific performance measures being tested, and to the tabular and graphic outputs used for the options. Examples of some of the adjustments that were made are described below.

Revision to "Basic" Asset Management Option

The definition of the "basic" asset management option was revised to focus on the fact that this option is based on defining renewal and replacement estimates for assets based on their estimated service lives. Use of "depreciation," which had been used to describe this option, was thought to be confusing because this option's primary purpose is in using useful service life, which in many cases is different than the factors that have traditionally been used to depreciate assets in fixed asset systems. Also, this option applies the service lives to the replacement value for the assets, whereas most traditional depreciation analyses apply factors to the original costs of the assets.

While the methodology for the basic option in some respects emulates the methods used in calculating depreciation for financial reporting purposes, the basic option is not a depreciation option per se. That is, the basic option computes renewal and replacement requirements by applying the expected service life factors to the replacement value of the assets; the basic option is *not* based on applying accounting-based factors to the original cost of the assets, as is done in a traditional depreciation analysis for financial-reporting purposes. Several of the participating utilities indicated that when defining "service life," they tended to consider not only the number of years that an asset can be used before it is physically worn out but also when it no longer serves a purpose. For example, a functioning pump may be changed out due to an overall upgrade to meet new regulations; or, it might be changed out because the cost to a utility of "manufacturing" replacement parts that are no longer available from the original supplier becomes prohibitively expensive. Another example is that a computer may still work but may be too slow or have too little memory or storage capacity to run software now required.

Revisions to Performance Measures

The "service reliability—main breaks" performance measure was revised from a measure of the raw number of breaks to the number of breaks per foot of pipe segment, with a revised scale developed that relates the number of breaks to the length of piping.

"R&R status" identifies whether previous interventions had been performed on the asset in the preceding 2 years. This measure was revised to *exclude* the full replacement of assets. That is, the revised question for this measure was formulated as, "Was the asset relined, rebuilt, or protected during the past two years?"

With the benefit of seeing how it was applied in the first round sample reports and figures, it was decided to remove "planned and reactive maintenance" and replace it with a new performance measure, "maintenance expenditures as a percent of asset replacement value." A scale that assigns points to each asset on the basis of the resulting percentage was developed. The data inputs for this measure were revised to include reactive maintenance, defined as immediate response to asset failure, and preventive maintenance, which was defined for the purpose of this effort as all other maintenance activities (but *not* full asset replacement).

Addition of Outreach Graphics

The research team added several figures to the model after the second project workshop to illustrate the kinds of graphic summaries that might be useful in communicating the findings and recommendations of the asset management options to decision-making boards and stakeholder groups. This was done in response to a facilitated discussion at the second project workshop meeting about audiences for the asset management. During that discussion, several participants noted that though many of the tables and figures in the template model were necessarily complex to facilitate the "command-and-control" scenario analysis that provides significant value to utility staffs in identifying and evaluating candidate assets or groups of assets for renewal or replacement action, they are because of this complexity not ideal for communicating findings and recommendations to outside groups, which are neither as heavily immersed in the technical details nor as interested in the details of the analytical process behind the findings. Thus for the final version of the template model, the research team added several graphics to illustrate how the findings and recommendations of the options studied might be conveyed graphically to decision-making bodies and other external stakeholder groups.

DEVELOPMENT OF EVALUATION FRAMEWORK

Evaluating the asset management options studied is considered one of the key elements of this project. The approach called for providing evaluation through four primary vehicles:

- Findings of the research team, based on review of the outputs of the study in relation to the team's experience with asset management issues in other contexts
- Scorecards that would be completed by each of the participating utilities to "grade" the value of the options, parameters, and outputs studied and to assess the difficulty of developing and maintaining the information required to support the various options studied
- Letters of comment by each utility, providing additional context and insights based on the experience in developing the input data and reviewing the outputs of the model.
- Compilation and summary of the utility evaluations (scorecards and letters of comment)

Development of the evaluation framework began early in the project, in order to allow input on the framework by the utilities. The research team developed a working draft of the scorecard and reviewed it with the utility participants at the second workshop meeting. The scorecard includes separate assessments of the usefulness/value of the options, performance measures, and outputs to aid management decision-making and to aid in communicating asset management needs to decision-making bodies and other external stakeholders. Breakout-group discussions were used to provide feedback on the draft scorecard.

The scorecard was revised in response to the comments received at the second workshop. For example, several additional questions were added to the score card, including "What level of planning are you [the participant] going to implement?" and "Did you have difficulty using data from multiple databases?"

At the third workshop meeting, small breakout groups were used to identify specific internal and external stakeholders to the asset management process for each participating utility, to help focus their evaluations on likely specific audiences. The wide range of audiences is summarized in Chapter 7, in the section "Audiences for Asset Management Planning Information."

A sample of the final scorecard that was completed by the participating utilities can be found beginning on p. 15 (Figure 2.1). The completed scorecards for each utility are include in the "Participating Utilities' Evaluations and Sample Data Sets" section of the appendix.

Table 2.1
Performance measure lists from first project workshop meeting

Effectiveness	Efficiency	Quality
Number of main breaks/ length	Treatment cost	Customer complaints (taste, odor)
Number or outages (interruptions)	Maintenance cost	Response time to complaint's
PSI for fire flow	Rate increases	Time to solve a problem
Number of shut downs of primary source water	Full time employees	Water quality at tap
Percentage of well fields available	Effectiveness of information systems	Compliance (number of violations
Percent of assets available for service	Integrated processes	Safety
Miles of pipe replaced per year	Emergency response time	Lost time due to safety
Failure of critical service components	Number of spills	Customer safety
Duration of downtime (number of people impacted)	Percent of assets in PM system	Raw water quality
Miles of forest road out of service	Replacement of substandard services per year	Water conservation by sector
—	Unaccounted for water	Percent of customers participating in discount program
—	Preventive maintenance cost per linear foot	Miles of stream restored
—	Cost to treat per acre-foot	Turbidity
—	Leakage of source water supply	Flow monitoring
—	Planned versus unplanned outages to customers	Waterborne disease outbreaks
—	Age relative to life cycle	Number of customers participating in water conservation
—	Asset condition index versus target	Nutrient loading
—	Percent of preventive versus reactive maintenance	—
—	Percent of money in ground	—
—	Cost per family below EPA standard	—

Scorecard For Rating Asset Management Options

COVER SHEET

Utility Name: ________________________________

Number of service connections: _________

Population served: ___________

Type of Utility (check all that apply): Wholesaler ____ **Retailer**____

Services provided (check all that apply):

Water Treatment____ **Water Transmission/Distribution**____

Wastewater Treatment____ **Wastewater Collection**____

Stormwater____ **Other (describe)** ______________________

Please fill out the scorecard to the best of your ability, and please attach your partner letter of comment. If you feel a particular measurement does not apply to the category you are scoring, please just leave it blank. Your participation in this evaluation is an essential component of the project—thank you for taking the time to complete it.

Figure 2.1 Scorecard for rating asset management options

Scorecard For Rating Asset Management Options

Performance Measures—value to management (typically for asset hierarchy level 3 or 4)

How valuable is this information to management decision-making?	Not very valuable		Somewhat valuable		Very valuable
	1	2	3	4	5
Service reliability—interruptions (plant)					
Service reliability—interruptions (distribution system)					
Service reliability—main breaks (distribution system)					
Water quality (plant)					
Water quality (distribution system)					
R&R status (plant)					
R&R status (distribution system)					
Maintenance activity (plant)					
Maintenance activity (distribution system)					
Preventive maintenance (plant)					
Preventive maintenance (distribution system)					
Age vs. service life (plant)					
Age vs. service life (distribution system)					

Figure 2.1 (Continued)

Scorecard For Rating Asset Management Options

Performance Measures—value to stakeholders (typically for asset hierarchy level 1 or 2)

How valuable is this information in "making the case" to governing boards and stakeholders?	Not very valuable 1	2	Somewhat valuable 3	4	Very valuable 5
Service reliability—interruptions (plant)					
Service reliability—interruptions (distribution system)					
Service reliability—main breaks (distribution system)					
Water quality (plant)					
Water quality (distribution system)					
R&R status (plant)					
R&R status (distribution system)					
Maintenance activity (plant)					
Maintenance activity (distribution system)					
Preventive maintenance (plant)					
Preventive maintenance (distribution system)					
Age vs. service life (plant)					
Age vs. service life (distribution system)					

Figure 2.1 (Continued)

Scorecard For Rating Asset Management Options

Performance Measures—degree of difficulty

How difficult was it to obtain this information?	Very difficult		Somewhat difficult		Not very difficult
	1	2	3	4	5
Service reliability—interruptions (plant)					
Service reliability—interruptions (distribution system)					
Service reliability—main breaks (distribution system)					
Water quality (plant)					
Water quality (distribution system)					
R&R status (plant)					
R&R status (distribution system)					
Maintenance activity (plant)					
Maintenance activity (distribution system)					
Preventive maintenance (plant)					
Preventive maintenance (distribution system)					
Age vs. service life (plant)					
Age vs. service life (distribution system)					

Figure 2.1 (Continued)

Scorecard For Rating Asset Management Options

Asset Management Options

How would you rate the overall value of this option (for plant assets)?	Not very valuable		Somewhat valuable		Very valuable
	1	2	3	4	5
Basic Option (based on service lives)					
High-end Option (performance measures)					
Strategic Option (performance measures with time dimension)					
How would you rate the overall value of this option (for distribution system assets)?	**Not very valuable**		**Somewhat valuable**		**Very valuable**
	1	2	3	4	5
Basic Option (based on service lives)					
High-end Option (performance measures)					
Strategic Option (performance measures with time dimension)					

Figure 2.1 (Continued)

Scorecard For Rating Asset Management Options

Asset Management Options

How would you rate the data management challenges, e.g. electronic availability, hardware/software requirements (for plant assets)?	Very challenging		Somewhat challenging		Not overly challenging
	1	2	3	4	5
Basic Option (based on service lives)					
High-end Option (performance measures)					
Strategic Option (performance measures with time dimension)					

How would you rate the data management challenges, e.g. electronic availability, hardware/software requirements (for distribution system assets)?	Very challenging		Somewhat challenging		Not overly challenging
	1	2	3	4	5
Basic Option (based on service lives)					
High-end Option (performance measures)					
Strategic Option (performance measures with time dimension)					

Figure 2.1 (Continued)

Scorecard For Rating Asset Management Options

Asset Management Options

How would you rate the level of organizational and cultural challenges (for plant assets)?	**Many challenges**		**Some challenges**		**Not many challenges**
	1	**2**	**3**	**4**	**5**
Basic Option (based on service lives)					
High-end Option (performance measures)					
Strategic Option (performance measures with time dimension)					
How would you rate the level of organizational and cultural challenges (for distribution system assets)?	**Many challenges**		**Some challenges**		**Not many challenges**
	1	**2**	**3**	**4**	**5**
Basic Option (based on service lives)					
High-end Option (performance measures)					
Strategic Option (performance measures with time dimension)					

Figure 2.1 (Continued)

Scorecard For Rating Asset Management Options

Asset Management Options

How would you rate the value of this option to management decision-making (for plant assets)?	**Not very valuable**		**Somewhat valuable**		**Very valuable**
	1	**2**	**3**	**4**	**5**
Basic Option (based on service lives)					
High-end Option (performance measures)					
Strategic Option (performance measures with time dimension)					
How would you rate the value of this option to management decision-making (for distribution system assets)?	**Not very valuable**		**Somewhat valuable**		**Very valuable**
	1	**2**	**3**	**4**	**5**
Basic Option (based on service lives)					
High-end Option (performance measures)					
Strategic Option (performance measures with time dimension)					
How would you rate the value of this option to “making the case” to governing boards and stakeholders (for plant assets)?	**Not very valuable**		**Somewhat valuable**		**Very valuable**
	1	**2**	**3**	**4**	**5**
Basic Option (based on service lives)					
High-end Option (performance measures)					
Strategic Option (performance measures with time dimension)					
How would you rate the value of this option to “making the case” to governing boards and stakeholders (for distribution system assets)?	**Not very valuable**		**Somewhat valuable**		**Very valuable**
	1	**2**	**3**	**4**	**5**
Basic Option (based on service lives)					
High-end Option (performance measures)					
Strategic Option (performance measures with time dimension)					

Figure 2.1 (Continued)

Scorecard For Rating Asset Management Options

Reporting Outputs

How valuable is this information to management decision-making (for plant assets)?	Not very valuable		Somewhat valuable		Very valuable
	1	2	3	4	5
Tables					
B-1 Summarizes asset replacement cost, service life and remaining value for all asset classes					
B-2 Summarizes target R&R and maintenance activity for all asset classes					
B-3 Reports asset replacement cost, service life used and remaining value for individual assets					
H-1 Summarizes weighted performance score, by criteria, for all asset classes					
H-2 Reports raw performance score, by criteria, for individual assets					
S-1 Platform that enables multi-year capital expenditure and R&R planning across individual assets					
Figures					
B-1 Summarizes % service life used and remaining for asset class or individual assets					
B-2 Compares historical maintenance activity to target R&R for asset class or individual assets					
H-1 Presents performance score, by component, for asset class and individual assets					
H-2 Displays value allocation across the Asset Condition Rating (ACR) for asset classes					
S-1 Demonstrates impacts of capital expenditure and R&R planning on net system value over a 20-year period					
S-2 Graphs proposed capital and R&R expenditures across a 20-year forecast period					
Outreach Graphics					
O-1 Graphs service life used by major system component					
O-2 Displays overall performance score for the entire system and by major system component					
O-3 Presents Asset Condition Rating (ACR) by major system component					
O-4 Demonstrates impact of proposed capital + R&R programs on net system value					

Figure 2.1 (Continued)

Scorecard For Rating Asset Management Options

Reporting Outputs

How valuable is this information to management decision-making (for distribution system assets)?	Not very valuable		Somewhat valuable		Very valuable
	1	2	3	4	5
Tables					
B-1 Summarizes asset replacement cost, service life and remaining value for all asset classes					
B-2 Summarizes target R&R and maintenance activity for all asset classes					
B-3 Reports asset replacement cost, service life used and remaining value for individual assets					
H-1 Summarizes weighted performance score, by criteria, for all asset classes					
H-2 Reports raw performance score, by criteria, for individual assets					
S-1 Platform that enables multi-year capital expenditure and R&R planning across individual assets					
Figures					
B-1 Summarizes % service life used and remaining for asset class or individual assets					
B-2 Compares historical maintenance activity to target R&R for asset class or individual assets					
H-1 Presents performance score, by component, for asset class and individual assets					
H-2 Displays value allocation across the Asset Condition Rating (ACR) for asset classes					
S-1 Demonstrates impacts of capital expenditure and R&R planning on net system value over a 20-year period					
S-2 Graphs proposed capital and R&R expenditures across a 20-year forecast period					
Outreach Graphics					
O-1 Graphs service life used by major system component					
O-2 Displays overall performance score for the entire system and by major system component					
O-3 Presents Asset Condition Rating (ACR) by major system component					
O-4 Demonstrates impact of proposed capital + R&R programs on net system value					

Figure 2.1 (Continued)

Scorecard For Rating Asset Management Options

Reporting Outputs

How valuable is this information to "making the case" to governing boards and stakeholders (for plant assets)?	Not very valuable		Somewhat valuable		Very valuable
	1	2	3	4	5
Tables					
B-1 Summarizes asset replacement cost, service life and remaining value for all asset classes					
B-2 Summarizes target R&R and maintenance activity for all asset classes					
B-3 Reports asset replacement cost, service life used and remaining value for individual assets					
H-1 Summarizes weighted performance score, by criteria, for all asset classes					
H-2 Reports raw performance score, by criteria, for individual assets					
S-1 Platform that enables multi-year capital expenditure and R&R planning across individual assets					
Figures					
B-1 Summarizes % service life used and remaining for asset class or individual assets					
B-2 Compares historical maintenance activity to target R&R for asset class or individual assets					
H-1 Presents performance score, by component, for asset class and individual assets					
H-2 Displays value allocation across the Asset Condition Rating (ACR) for asset classes					
S-1 Demonstrates impacts of capital expenditure and R&R planning on net system value over a 20-year period					
S-2 Graphs proposed capital and R&R expenditures across a 20-year forecast period					
Outreach Graphics					
O-1 Graphs service life used by major system component					
O-2 Displays overall performance score for the entire system and by major system component					
O-3 Presents Asset Condition Rating (ACR) by major system component					
O-4 Demonstrates impact of proposed capital + R&R programs on net system value					

Figure 2.1 (Continued)

Scorecard For Rating Asset Management Options

Reporting Outputs

How valuable is this information to "making the case" to governing boards and stakeholders (for distribution system assets)?	Not very valuable		Somewhat valuable		Very valuable
	1	2	3	4	5
Tables					
B-1 Summarizes asset replacement cost, service life and remaining value for all asset classes					
B-2 Summarizes target R&R and maintenance activity for all asset classes					
B-3 Reports asset replacement cost, service life used and remaining value for individual assets					
H-1 Summarizes weighted performance score, by criteria, for all asset classes					
H-2 Reports raw performance score, by criteria, for individual assets					
S-1 Platform that enables multi-year capital expenditure and R&R planning across individual assets					
Figures					
B-1 Summarizes % service life used and remaining for asset class or individual assets					
B-2 Compares historical maintenance activity to target R&R for asset class or individual assets					
H-1 Presents performance score, by component, for asset class and individual assets					
H-2 Displays value allocation across the Asset Condition Rating (ACR) for asset classes					
S-1 Demonstrates impacts of capital expenditure and R&R planning on net system value over a 20-year period					
S-2 Graphs proposed capital and R&R expenditures across a 20-year forecast period					
Outreach Graphics					
O-1 Graphs service life used by major system component					
O-2 Displays overall performance score for the entire system and by major system component					
O-3 Presents Asset Condition Rating (ACR) by major system component					
O-4 Demonstrates impact of proposed capital + R&R programs on net system value					

Figure 2.1 (Continued)

Scorecard For Rating Asset Management Options

Degree of Difficulty

How would you rate the level of difficulty of using data from multiple databases (for plant assets)?	Very difficult		Somewhat difficult		Not very difficult
	1	2	3	4	5

How would you rate the level of difficulty of using data from multiple databases (for distribution system assets)?	Very difficult		Somewhat difficult		Not very difficult
	1	2	3	4	5

Planned Level of Implementation

What level of planning are you going to implement (for plant assets)?	Basic Option (based on service lives)	High-end Option (performance measures)	Strategic Option (performance measures w/time dimension)

What level of planning are you going to implement (for distribution system assets)?	Basic Option (based on service lives)	High-end Option (performance measures)	Strategic Option (performance measures w/time dimension)

Figure 2.1 (Continued)

CHAPTER 3
ASSET MANAGEMENT OPTIONS SELECTED FOR STUDY

BACKGROUND TO THE DEFINITION OF OPTIONS

Defining the appropriate options to study during this project was critical. It was deemed important to provide a reasonable range of options to reflect utilities' different levels of access to information, degree of commitment of resources, management, and stakeholder interest in asset management. It was also deemed important to define a range of options that reflects real-world conditions; that is, defining a high-end option that is unrealistic in terms of assumed information and resources would provide little value to the water industry in moving forward the industry's consideration and implementation of asset management programs.

As a result of these considerations, the definition of the options was done in a collaborative fashion during the first 6 months of the process, during which the participating utilities were substantially engaged. The Research Team defined three options (see following page for definitions) and then used the project workshops and other project communications to refine and adjust the options in response to comments from the participating utilities and AwwaRF's Project Advisory Committee. Examples of the refinements that took place include the following:

- The Basic Option has been redefined as a "service-life-based" option, rather than using depreciation terminology to define it, in order to distinguish this option from depreciation analyses, usually based on original costs depreciated, which is typically used in a utility's fixed asset system and financial audit reports.
- The High-End and Strategic Options are based on the use of performance measures. The specific performance measures worth studying and for which the participating utilities had a reasonable opportunity to gather performance information was defined cooperatively with the participating utilities during project workshops, monthly conference calls, and other communications. ("Performance measure" and "performance criteria" are used interchangeably in this study.)
- The Strategic Option initially was intended to include a risk analysis element. Because of the substantial data-gathering challenges that were identified in gathering the basic asset background, financial, and performance information, it became clear that many of the utilities would not be able to gather or provide the inputs for a risk analysis for their data sets within the time limitations of the project. This element of the Strategic Option was therefore removed. Several of the utilities that had the requisite information available did include "criticality" and "consequence" factors as supplemental performance factors.

Although most of the refinements to the options took place during the first 6 months of the project, some refinements were ongoing. For example, some of the performance measures and reporting outputs for the options were refined in response to ongoing input from the participating utilities during most of the project.

This chapter provides an overview of the three options, as finally defined and used in this study. In addition, each of the tables and figures developed for the options is described and illustrations of some of the key figures and tables are included. Samples of the outputs for each of the 11 utilities are included in the appendix; those samples show how the sample reports might look for utilities of varying sizes and organizational structures and organize data into differing types of hierarchies. See Chapter 6 for a more detailed discussion of the several levels of hierarchy that were defined for the sample data sets for this project.

OVERVIEW OF THE OPTIONS

In order to test the viability and usefulness of several levels of asset management planning and reporting, the following three options were developed and evaluated by each of the 11 participating utilities:

- **Basic**, in which traditional concepts of valuation and service life are used to approximate condition, estimate renewal and replacement factors, and help to target renewal expenditures. Key concepts used in this option are the replacement value of the assets and realistic estimates of the useful lives of the assets.
- **High-End**, in which key performance measures such as service outages and regulatory compliance are identified and prioritized. Performance scores are then used to assess the condition of individual and grouped assets to help guide renewal and replacement decision-making.
- **Strategic**, which incorporates elements of the basic and high-end options, but uses additional features to estimate the financial impacts of renewal and replacement decision-making and its effect on system value and/or condition over time.

The "Participating Utilities' Evaluations and Sample Data Sets" portion of the appendix includes at least one page of sample output for each of the tables and figures of the 11 studied systems. In this chapter, examples from the Santa Clara Valley Water District (SCVWD) are used to illustrate the tables and figures.[1] As detailed in the descriptions of the tables and figures below, in some cases multiple examples are provided for some of the figures in order to illustrate the various levels at which comparisons can be made.

AN IMPORTANT NOTE ABOUT THE EXAMPLES INCLUDED: The examples shown in this chapter and in the appendix are based on a limited but real sample of the assets for the participating utilities. The scores that are shown should NOT *be considered final evaluations of the overall condition or renewal and replacement needs of the 11 systems studied.*

- *In some cases, the cost, background, and performance information is based on very reliable "final" information about the assets. In other cases, the information is based on best-available estimates, which in some cases are considered quite reliable and in other cases less so. In some cases, the Participating Utilities were simply not able to complete information for some of the measures due to data availability issues. This simulates the real-world situation faced by almost all utilities establishing asset*

[1] To aid in cross-referencing, tables and figures in this chapter which are part of the "Reporting Samples" section for each participating utility in the Appendix retain the numbering system used there (e.g., Table B-1, Figure H-2), in which "B" denotes "basic," "H" denotes "high-end," "S" denotes "strategic," and "O" denotes "outreach." All other tables and figures in this report follow the standard AwwaRF numbering scheme (e.g., Table 3.1 is the first table in this chapter that is not part of the "Reporting Samples" in the Appendix).

management programs (i.e., the need to work with a variety of levels of information and completeness of information).

- *The participating utilities in some cases selected assets or asset groups that would illustrate contrast in the reports, so the asset groups selected for study may not be representative of the age or condition characteristics of the systems as a whole. Also, because the study was limited to 1,000-record samples, the results are considered representative for the assets selected but are* NOT *necessarily an overall picture of the status and needs of the systems as a whole.*
- *The participating utilities did not have the opportunity to view multiple iterations of weighting and scoring before the development of the sample tables and figures included in the report. Their weighting and scaling of the performance measures would, in some cases, likely be refined through a process of such testing. Some of the utilities have since secured copies of the template model used to develop the report, and have performed sensitivity analyses related to criteria weighting and performance scaling. Some of these utilities have commented on the results of these efforts as part of their letters of comment, which are included in the "Participating Utilities' Evaluations and Sample Data Sets" section of the appendix.*

BASIC OPTION

The several tables and figures developed for the basic option focus on the value of the system and the percent expended, based on realistic estimates of the useful lives of the assets. A challenge for this option is to define real-world useful life factors that take into account both industry experience and a utility's specific context (e.g., soil and climate conditions, maintenance practices). The tables and figures are listed below, and examples follow:

- **Table B-1: Service Life Used** presents a summary (in outline form) of replacement cost, service life used, and remaining value across all levels of a utility's asset hierarchy.
- **Table B-2: Target R&R** summarizes (in outline form) preventive and reactive maintenance by asset class, and identifies target renewal and replacement spending—the level of annual reinvestment required to maintain system value.
- **Table B-3: Replacement Cost and Asset Value** reports replacement cost, basic service life information, and target R&R for individual assets of the system.[2]
- **Figure B-1: Service Life Analysis** shows remaining value and service life used across the asset hierarchy in a graphical format. Information can be compared across the same level (plant vs. distribution assets, Segment A vs. Segment B) or at different levels (distribution assets overall vs. Segment A) to contrast various states of remaining value.[3] For example, the top figure (p. 35) shows a comparison of the SCVWD's distribution system assets studies vs. the portion of the District's plant

[2] The utility participants developed data sets of up to 1,000 records. Since printing the entire listing of assets for most of the sample files is not practical, the report will include one representative sheet for this table and the other tables that are asset-specific.

[3] Several examples of this figure are included in the report appendix to illustrate the options for using this figure, where supported by the utilities' asset hierarchies.

assets included in the study. The lower figure on that page shows a comparison of the distribution system overall to the Central Pipeline, a "level 2" asset.

- **Figure B-2: Maintenance Activity**, charts preventive and reactive maintenance against target R&R. As with Figure B-1, information can be compared across multiple levels of the asset hierarchy.[4] For example, the top figure (p. 36) shows a comparison of the SCVWD's distribution system assets studied vs. the portion of the District's plant assets included in the study. The bottom figure compares the distribution system overall to Almaden Valley Pipeline, a level 2 asset.

Table B-1
Service life used

Asset Management Planning and Reporting Options (AMPRO)
Santa Clara Valley Water District

PRINT Summary

Choose Data Set
(●) Service Life Used ( ) Performance Scores
( ) Target R&R ( ) Show All Data Sets

Table B-1: SERVICE LIFE USED

	Replacement Cost	Service Life Used	Remaining Value	% Used	% Remaining
TOTAL SYSTEM	**$ 120,348,547**	**$ 36,304,631**	**$ 84,043,916**	**30.2%**	**69.8%**
SUMMARY, Asset Hierarchy					
Distribution System	**$ 119,883,467**	**$ 35,926,965**	**$ 83,956,502**	**30.0%**	**70.0%**
Almaden Valley Pipeline	*$ 53,290,635*	*$ 15,533,068*	*$ 37,757,567*	*29.1%*	*70.9%*
AVP Vasona - Guadalupe	$ 19,208,631	$ 7,732,866	$ 11,475,765	40.3%	59.7%
AVP Guadalupe - Shannon	$ 18,741,424	$ 4,291,083	$ 14,450,341	22.9%	77.1%
AVP Shannon - Calero	$ 15,340,580	$ 3,509,119	$ 11,831,461	22.9%	77.1%
Snell Pipeline	*$ 26,367,011*	*$ 4,021,058*	*$ 22,345,953*	*15.3%*	*84.7%*
SPL Colleen - Branham	$ 9,692,977	$ 1,476,255	$ 8,216,722	15.2%	84.8%
SPL Branham - East Pipeline	$ 16,674,034	$ 2,544,803	$ 14,129,231	15.3%	84.7%
Central Pipeline	*$ 40,225,821*	*$ 16,372,839*	*$ 23,852,982*	*40.7%*	*59.3%*
CPL Piedmont Yard to Guadalupe Turnout	$ 16,433,681	$ 6,798,109	$ 9,635,572	41.4%	58.6%
CPL Guadalupe Turnout to Vasona Pump Station	$ 23,792,140	$ 9,574,730	$ 14,217,410	40.2%	59.8%
Plant Assets	**$ 465,080**	**$ 377,667**	**$ 87,413**	**81.2%**	**18.8%**
RWTP Chlorine (OCL) System	*$ 176,170*	*$ 149,843*	*$ 26,327*	*85.1%*	*14.9%*
RWTP OCL Storage Tank # 1	$ 50,890	$ 46,837	$ 4,053	92.0%	8.0%
RWTP OCL Storage Tank # 2	$ 50,890	$ 46,837	$ 4,053	92.0%	8.0%
RWTP OCL Storage Tank # 3	$ 50,890	$ 46,103	$ 4,787	90.6%	9.4%
RWTP OCL Room 224	$ 23,500	$ 10,067	$ 13,433	42.8%	57.2%
PWTP Chlorine (OCL) System	*$ 122,530*	*$ 105,737*	*$ 16,793*	*86.3%*	*13.7%*
PWTP OCL Room	$ 23,500	$ 10,067	$ 13,433	42.8%	57.2%
PWTP OCL Storage Tank # 1	$ 32,677	$ 31,557	$ 1,120	96.6%	3.4%
PWTP OCL Storage Tank # 2	$ 32,677	$ 31,557	$ 1,120	96.6%	3.4%
PWTP OCL Storage Tank # 3	$ 32,677	$ 31,557	$ 1,120	96.6%	3.4%
PWTP OCL Storage Sump Pump	$ 1,000	$ 1,000	$ -	100.0%	0.0%
STWTP Chlorine (OCL) System	*$ 166,380*	*$ 122,087*	*$ 44,293*	*73.4%*	*26.6%*
STWTP Hypochlorite Feed Room	$ 58,000	$ 26,967	$ 31,033	46.5%	53.5%
STWTP Hypochlorite Storage Area	$ 108,380	$ 95,120	$ 13,260	87.8%	12.2%
Check	**120,348,547**	**$ 36,304,631**	**$ 84,043,916**		

[4] Several examples of this figure are included in the report appendix to illustrate the options for using this figure, where supported by the utilities' asset hierarchies.

Table B-2
Target R&R

Asset Management Planning and Reporting Options (AMPRO)
Santa Clara Valley Water District

PRINT Summary

Choose Data Set
○ Service Life Used ○ Performance Scores
◉ Target R&R ○ Show All Data Sets

Table B-2: TARGET R&R	Target R&R	Preventive Maintenance	Reactive Maintenance	Reactive, % of Preventive
TOTAL SYSTEM	**$ 1,378,677**	**$ 85,176**	**$ 394,570**	**463.2%**
SUMMARY, Asset Hierarchy				
Distribution System	**$ 1,306,181**	**$ 69,592**	**$ 271,308**	**389.9%**
Almaden Valley Pipeline	*$ 580,515*	*$ 24,883*	*$ 48,060*	*193.1%*
AVP Vasona - Guadalupe	$ 209,077	$ 11,564	$ 20,679	178.8%
AVP Guadalupe - Shannon	$ 204,337	$ 8,985	$ 19,837	220.8%
AVP Shannon - Calero	$ 167,101	$ 4,334	$ 7,544	174.1%
Snell Pipeline	*$ 287,218*	*$ 13,460*	*$ 77,740*	*577.5%*
SPL Colleen - Branham	$ 105,447	$ 2,259	$ 23,337	1033.1%
SPL Branham - East Pipeline	$ 181,772	$ 11,201	$ 54,403	485.7%
Central Pipeline	*$ 438,447*	*$ 31,249*	*$ 145,508*	*465.6%*
CPL Piedmont Yard to Guadalupe Turnout	$ 178,940	$ 12,933	$ 51,185	395.8%
CPL Guadalupe Turnout to Vasona Pump Station	$ 259,507	$ 18,316	$ 94,323	515.0%
Plant Assets	**$ 72,496**	**$ 15,584**	**$ 123,262**	**790.9%**
RWTP Chlorine (OCL) System	*$ 29,061*	*$ 1,787*	*$ 54,437*	*3046.0%*
RWTP OCL Storage Tank # 1	$ 9,165	$ 357	$ 14,484	4056.3%
RWTP OCL Storage Tank # 2	$ 9,165	$ 357	$ 13,667	3827.5%
RWTP OCL Storage Tank # 3	$ 9,165	$ 226	$ 19,587	8671.1%
RWTP OCL Room 224	$ 1,567	$ 847	$ 6,699	790.8%
PWTP Chlorine (OCL) System	*$ 20,533*	*$ 678*	*$ 31,910*	*4708.6%*
PWTP OCL Room	$ 1,567	$ 678	$ 1,560	230.2%
PWTP OCL Storage Tank # 1	$ 6,255	$ -	$ 10,286	0.0%
PWTP OCL Storage Tank # 2	$ 6,255	$ -	$ 10,032	0.0%
PWTP OCL Storage Tank # 3	$ 6,255	$ -	$ 10,032	0.0%
PWTP OCL Storage Sump Pump	$ 200	$ -	$ -	0.0%
STWTP Chlorine (OCL) System	*$ 22,903*	*$ 13,120*	*$ 36,915*	*281.4%*
STWTP Hypochlorite Feed Room	$ 4,267	$ 6,560	$ 4,193	63.9%
STWTP Hypochlorite Storage Area	$ 18,636	$ 6,560	$ 32,722	498.8%
Check	**$ 1,378,677**	**$ 85,176**	**$ 394,570**	

Table B-3
Replacement Cost and Asset Value

Asset Management Planning and Reporting Options (AMPRO)

Asset Replacement Cost & Performance Scores

SYSTEM ASSETS - HIERARCHY | **Table B-3: REPLACEMENT COST & ASSET VALUE**

Choose Data Set: (●) Table B-3, Replacement Cost & Asset Value ( ) Table H-2, Performance Scores (raw)

Level 1	Level 2	Level 3	Level 4	Escalation Index	Replacement Cost	TARGET R&R	Service Life Used	REMAINING VALUE
Distribution System	Central Pipeline	CPL Guadalupe Turnout t	CPL Piping - Sta. 666+09 to 666+47	6.99	$ 22,800	$ 248	$ 9,417	13,383
Plant Assets	RWTP Chlorine (OCL) Sy	RWTP OCL Storage Tank	RWTP OCL Storage Tank # 1	1.20	$ 42,290	$ 8,458	$ 42,290	-
Plant Assets	RWTP Chlorine (OCL) Sy	RWTP OCL Storage Tank	RWTP OCL Transfer Pump # 1	1.20	$ 2,100	$ 140	$ 980	1,120
Plant Assets	RWTP Chlorine (OCL) Sy	RWTP OCL Storage Tank	RWTP OCL Supply Pump # 1	1.20	$ 5,500	$ 367	$ 2,567	2,933
Plant Assets	RWTP Chlorine (OCL) Sy	RWTP OCL Storage Tank	RWTP OCL Sump Pump # 1 (storage)	1.20	$ 1,000	$ 200	$ 1,000	-
Plant Assets	RWTP Chlorine (OCL) Sy	RWTP OCL Storage Tank	RWTP OCL Storage Tank # 2	1.20	$ 42,290	$ 8,458	$ 42,290	-
Plant Assets	RWTP Chlorine (OCL) Sy	RWTP OCL Storage Tank	RWTP OCL Transfer Pump # 2	1.20	$ 2,100	$ 140	$ 980	1,120
Plant Assets	RWTP Chlorine (OCL) Sy	RWTP OCL Storage Tank	RWTP OCL Supply Pump # 2	1.20	$ 5,500	$ 367	$ 2,567	2,933
Plant Assets	RWTP Chlorine (OCL) Sy	RWTP OCL Storage Tank	RWTP OCL Sump Pump # 2 (storage)	1.20	$ 1,000	$ 200	$ 1,000	-
Plant Assets	RWTP Chlorine (OCL) Sy	RWTP OCL Storage Tank	RWTP OCL Storage Tank # 3	1.20	$ 42,290	$ 8,458	$ 42,290	-
Plant Assets	RWTP Chlorine (OCL) Sy	RWTP OCL Storage Tank	RWTP OCL Transfer Pump # 3	1.20	$ 2,100	$ 140	$ 980	1,120
Plant Assets	RWTP Chlorine (OCL) Sy	RWTP OCL Storage Tank	RWTP OCL Supply Pump # 3	1.12	$ 5,500	$ 367	$ 1,833	3,667
Plant Assets	RWTP Chlorine (OCL) Sy	RWTP OCL Storage Tank	RWTP OCL Sump Pump # 3 (storage)	1.20	$ 1,000	$ 200	$ 1,000	-
Plant Assets	RWTP Chlorine (OCL) Sy	RWTP OCL Room 224	RWTP OCL Exhaust Fan	1.20	$ 1,000	$ 67	$ 467	533
Plant Assets	RWTP Chlorine (OCL) Sy	RWTP OCL Room 224	RWTP OCL Meter Pump # 1	1.16	$ 4,500	$ 300	$ 1,800	2,700
Plant Assets	RWTP Chlorine (OCL) Sy	RWTP OCL Room 224	RWTP OCL Meter Pump # 2	1.16	$ 4,500	$ 300	$ 1,800	2,700
Plant Assets	RWTP Chlorine (OCL) Sy	RWTP OCL Room 224	RWTP OCL Meter Pump # 3	1.20	$ 4,500	$ 300	$ 2,100	2,400
Plant Assets	RWTP Chlorine (OCL) Sy	RWTP OCL Room 224	RWTP OCL Meter Pump # 4	1.16	$ 4,500	$ 300	$ 1,800	2,700
Plant Assets	RWTP Chlorine (OCL) Sy	RWTP OCL Room 224	RWTP OCL Meter Pump # 5	1.20	$ 4,500	$ 300	$ 2,100	2,400
Plant Assets	PWTP Chlorine (OCL) Sy	PWTP OCL Room	PWTP OCL Exhaust Fan	1.20	$ 1,000	$ 67	$ 467	533
Plant Assets	PWTP Chlorine (OCL) Sy	PWTP OCL Room	PWTP OCL Meter Pump # 1	1.20	$ 4,500	$ 300	$ 2,100	2,400
Plant Assets	PWTP Chlorine (OCL) Sy	PWTP OCL Room	PWTP OCL Meter Pump # 2	1.20	$ 4,500	$ 300	$ 2,100	2,400
Plant Assets	PWTP Chlorine (OCL) Sy	PWTP OCL Room	PWTP OCL Meter Pump # 3	1.20	$ 4,500	$ 300	$ 2,100	2,400
Plant Assets	PWTP Chlorine (OCL) Sy	PWTP OCL Room	PWTP OCL Meter Pump # 4	1.20	$ 4,500	$ 300	$ 2,100	2,400

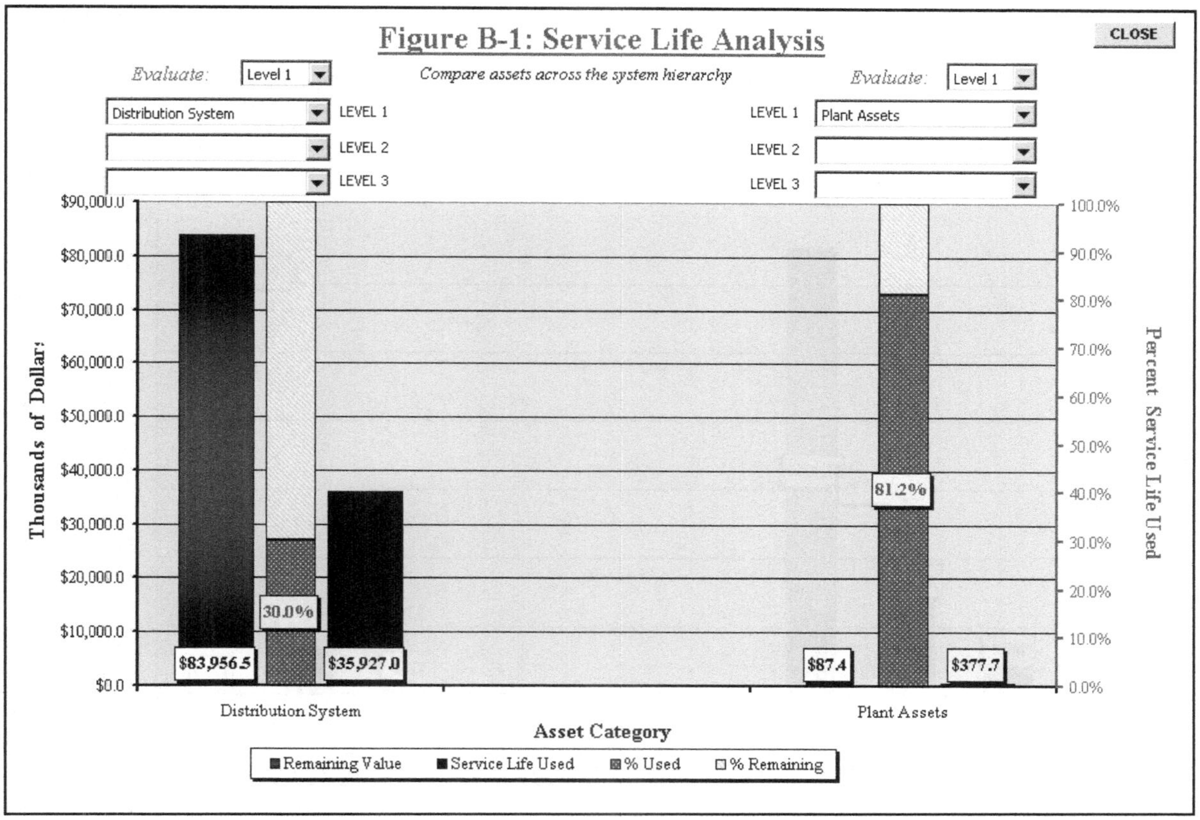

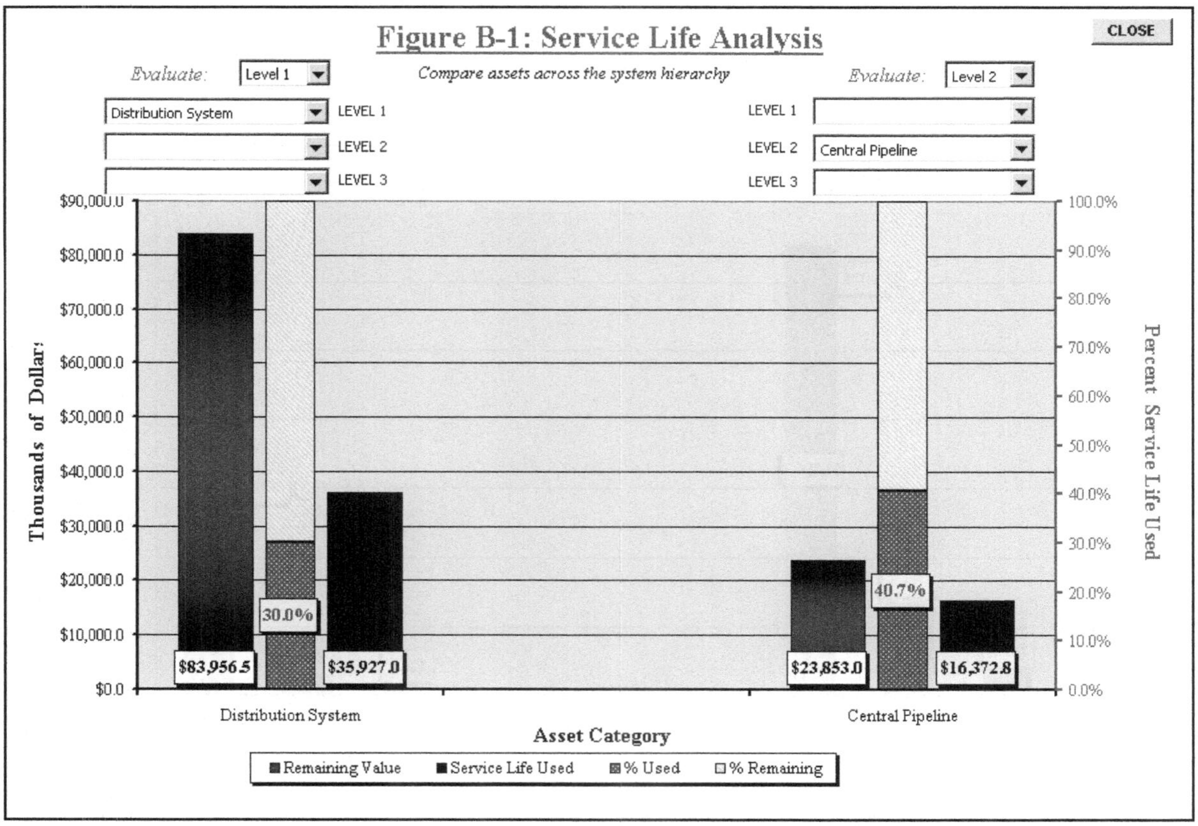

Figure B-1 Service Life Analysis (text on pages 31-32 details cases used in this illustration)

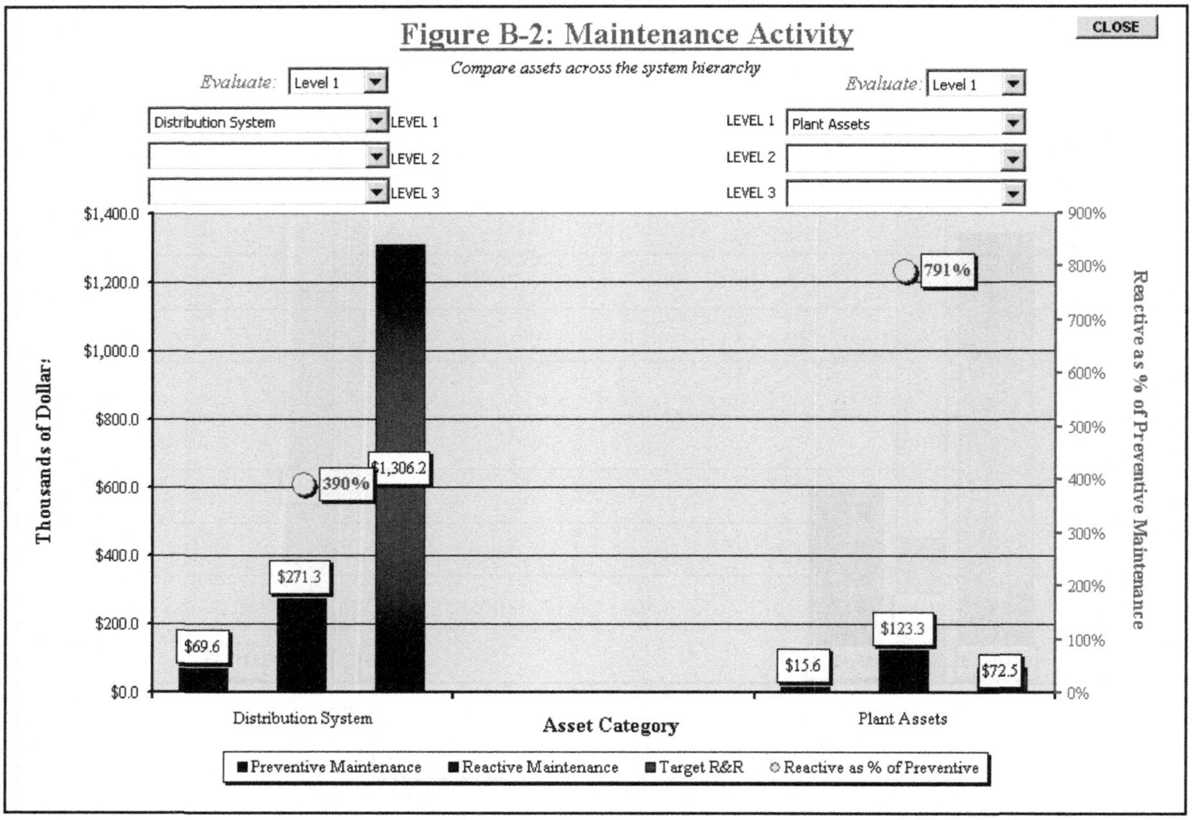

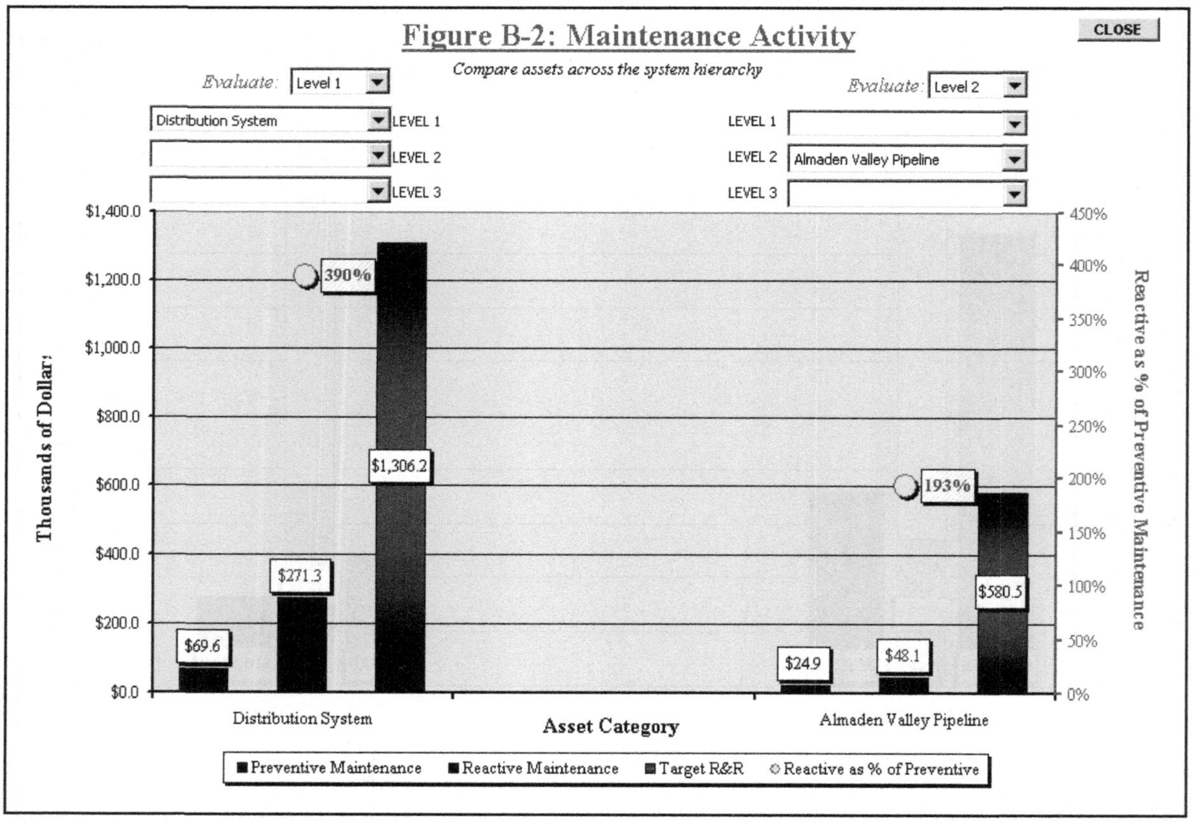

Figure B-2 Maintenance Activity (text on page 32 details cases used in this illustration)

HIGH-END OPTION

The several tables and figures developed for the high-end option focus on using key performance measures to provide insights regarding asset condition and renewal or replacement requirements. As detailed in Chapter 4, the seven performance measures for which data were gathered and evaluated by all 11 of the participating utilities are the following:

Table 3.1
Assumptions, condition rating framework, and performance criteria

Asset Management Planning and Reporting Options (AMPRO)
Santa Clara Valley Water District

Assumptions, Condition Rating Framework, and Performance Criteria

GENERAL ASSUMPTIONS

		Asset Condition Rating		Scores
Model Year	2003	**Poor**	0.0	6.5
Escalation	0.00%	**Good**	6.5	9.0
Calculation Check	Good	**Excellent**	9.0	10.0

PERFORMANCE CRITERIA & EMPHASIS

Criteria Weights: Distribution	Criteria Weights: Plant	* Partners can customize the scoring and relative weighting of each criterion	ADJUST Performance Scales	CLOSE Performance Scales	
0.30	0.30	**Service Reliability - Interruptions**	For Every	Deduct X	
		Number of Unplanned Service Interruptions	1	3	
0.30	0.00	**Service Reliability - Main Breaks**	Main Breaks Per Mile		Score
		Number of Main Breaks per Mile	0.00	0.00	10
			0.01	0.05	9
			0.06	0.10	5
			0.11	0.13	3
			0.14	MAX	0
0.15	0.25	**Water Quality**	For Every	Deduct X	
		Number of Violations, Internal Standards	1	4	
0.00	0.00	**R&R Status**	IF "No"..	IF "Yes"..	
		Asset relined, rebuilt, or protected?	4	10	
0.05	0.15	**Maintenance Activity**	Range of Percentages		Score
		Preventive & Reactive spending (combined) as % of replacement cost	0%	0%	8
			1%	10%	10
			11%	50%	7
			51%	75%	3
			76%	MAX	0
0.10	0.10	**Preventive Maintenance**		Possibilities	Score
		Was it Scheduled? Was money Spent? 100% Spent?		NNN	10
				YNN	0
				YYN	7
				YYY	10
0.10	0.20	**Age vs. Service Life**	Range of Percentages		Score
		Distribution Assets: Age as a % of Service Life	0%	50%	10
			51%	75%	6
			76%	100%	5
			101%	150%	2
			151%	MAX	0
		Plant Assets: Age as a % of Service Life	0%	40%	10
			41%	50%	8
			51%	75%	6
			76%	90%	3
			91%	MAX	0
1.00	**1.00**				

1. Service reliability—interruptions
2. Service reliability—main breaks (per mile of pipe)
3. Water quality
4. R&R status
5. Maintenance activity
6. Preventive maintenance
7. Age vs. service life

These performance measures were selected for study from 47 candidate measures, based on the project team's ranking of the importance of the candidates and the ability of the participating utilities to gather required data inputs during the several-month period allocated for data-gathering for the project. As also detailed in Chapter 4, in a few cases, the utilities have also gathered information for additional performance measures.

A performance score, or "scale," has been created in which each asset is given a score between 0 and 10 for each performance measure (10 is always defined as the best possible performance). In addition, the relative importance of the performance measures has been weighted by each utility, so that the relative importance of each of the performance measures is taken into consideration when calculating a final weighted score for each asset or group of assets. The participating utilities were also provided the option to customize the "scale" that is used to assign points for each of the performance measures. The chart above shows the weighting and scaling factors used by SCVWD for its test of the asset management options. The tables and figures are listed below; examples follow:

- **Table H-1: Performance Scores** summarizes (in outline form) the contribution of each performance measure to the overall performance score of asset levels within a utility system's hierarchy.
- **Table H-2: Performance Scores** presents the raw score (scaled from 0 to 10) for each performance criterion and the overall score for individual assets of the system.[1]
- **Figure H-1: Asset Performance** demonstrates how the score of each performance measure contributes to the overall performance score. As with other reporting figures, a comparison may be made at any of the three levels of the hierarchy. To provide a benchmark, the far left and right columns present the "perfect score" for distribution and plant assets, respectively.[2,5] The overall performance of the Total System is shown in the middle of the figure.[6] For the examples shown in this chapter for the SCVWD:
  - The first example (p.42, top) shows a comparison at the highest level (level 1), in which the overall performance of the distribution system assets studied is compared with the plant assets studied.
  - The second example (p. 42, bottom) shows the comparative performance of two level 2 portions of the asset hierarchy. In this case, a portion of the distribution

---

[5] Because of the nature of the performance criteria, different weightings were developed for plant and distribution assets.

[6] To avoid potential confusion that may result from the different weightings assigned to the distribution and plant systems, the Total System bar is not broken down by the contributing criteria.

system (the Central Pipeline), is compared with a part of the plant assets (the PWTP Chlorine OCL System).

  - The third example (p. 43) shows the comparative performance of two level 3 portions of the asset hierarchy. In this case, a subset of the level 2 central pipeline (SPL Colleen-Branham) is compared with a subset of the PWTP Chlorine OCL System (in this case, the SWTP OCL Storage Sump Pump).

- **Figure H-2: Condition Rating** allocates remaining value to one of three condition categories (poor, good, excellent) based on an index that relates performance score to condition.[2,7] The sample outputs for this figure (pp. 44 and 45) illustrate several of the levels of hierarchy for the SCVWD system. The top figure on page 44 illustrates outputs at the highest level (Level 1), in this case for all distribution system assets included in the analysis. The bottom figure on page 44 illustrates outputs at the second level of the hierarchy, in this case for the PWTP Chlorine (OCL) system. The figure on the top of page 45 illustrates outputs at the third level of hierarchy, in this case the STWTP Hypochlorite Feed Room.

---

[7] The asset condition rating index was developed using the following scoring: Poor (0–6.5), Good (6.5–9.0), and Excellent (9.0–10). This scoring was developed by the Research Team to impose a uniform scoring for the sample system data, since the participating utilities did not have the opportunity to view multiple iterations of weighting and scoring before the development of the sample tables and figures included in the report. Some of the utilities have since secured copies of the template model used to develop their system's data for this report, and have performed sensitivity analyses related to criteria weighting and performance scaling. Some of these utilities have commented on the results of these efforts as part of their letters of comment, which are included in the appendix.

Table H-1
Performance scores

Asset Management Planning and Reporting Options (AMPRO)
Santa Clara Valley Water District

PRINT Summary

Choose Data Set
- ○ Service Life Used
- ◉ Performance Scores
- ○ Target R&R
- ○ Show All Data Sets

Table H-1: PERFORMANCE SCORES

	Service Interruptions	Main Breaks	Water Quality	R&R Status	Maintenance Activity	Preventive Maintenance	Age vs. Service Life	TOTAL Score
TOTAL SYSTEM								**9.67**
SUMMARY, Asset Hierarchy								
Distribution System	**3.00**	**3.00**	**1.50**	**0.00**	**0.40**	**0.78**	**1.00**	**9.68**
Almaden Valley Pipeline	*3.00*	*3.00*	*1.50*	*0.00*	*0.40*	*1.00*	*1.00*	*9.90*
AVP Vasona - Guadalupe	3.00	3.00	1.50	0.00	0.40	1.00	1.00	9.90
AVP Guadalupe - Shannon	3.00	3.00	1.50	0.00	0.40	1.00	1.00	9.90
AVP Shannon - Calero	3.00	3.00	1.50	0.00	0.40	1.00	1.00	9.90
Snell Pipeline	*3.00*	*3.00*	*1.50*	*0.00*	*0.40*	*0.00*	*1.00*	*8.90*
SPL Colleen - Branham	3.00	3.00	1.50	0.00	0.40	0.00	1.00	8.90
SPL Branham - East Pipeline	3.00	3.00	1.50	0.00	0.40	0.00	1.00	8.90
Central Pipeline	*3.00*	*3.00*	*1.50*	*0.00*	*0.40*	*1.00*	*1.00*	*9.90*
CPL Piedmont Yard to Guadalupe Turnout	3.00	3.00	1.50	0.00	0.40	1.00	1.00	9.90
CPL Guadalupe Turnout to Vasona Pump Station	3.00	3.00	1.50	0.00	0.40	1.00	1.00	9.90
Plant Assets	**3.00**	**0.00**	**2.50**	**0.00**	**1.07**	**0.97**	**0.59**	**8.13**
RWTP Chlorine (OCL) System	*3.00*	*0.00*	*2.50*	*0.00*	*1.05*	*0.98*	*0.46*	*8.00*
RWTP OCL Storage Tank # 1	3.00	0.00	2.50	0.00	1.06	0.98	0.24	7.78
RWTP OCL Storage Tank # 2	3.00	0.00	2.50	0.00	1.05	0.98	0.24	7.77
RWTP OCL Storage Tank # 3	3.00	0.00	2.50	0.00	1.02	0.98	0.28	7.78
RWTP OCL Room 224	3.00	0.00	2.50	0.00	1.14	1.00	1.83	9.47
PWTP Chlorine (OCL) System	*3.00*	*0.00*	*2.50*	*0.00*	*1.13*	*0.93*	*0.40*	*7.96*
PWTP OCL Room	3.00	0.00	2.50	0.00	1.40	0.96	1.68	9.54
PWTP OCL Storage Tank # 1	3.00	0.00	2.50	0.00	1.06	0.94	0.10	7.60
PWTP OCL Storage Tank # 2	3.00	0.00	2.50	0.00	1.06	0.94	0.10	7.60
PWTP OCL Storage Tank # 3	3.00	0.00	2.50	0.00	1.06	0.94	0.10	7.60
PWTP OCL Storage Sump Pump	3.00	0.00	2.50	0.00	1.20	0.00	0.00	6.70
STWTP Chlorine (OCL) System	*3.00*	*0.00*	*2.50*	*0.00*	*1.06*	*0.99*	*0.85*	*8.40*
STWTP Hypochlorite Feed Room	3.00	0.00	2.50	0.00	1.09	0.98	1.69	9.26
STWTP Hypochlorite Storage Area	3.00	0.00	2.50	0.00	1.04	1.00	0.40	7.94
Check								

Table H-2
Performance scores

Asset Management Planning and Reporting Options (AMPRO)

Asset Replacement Cost & Performance Scores

SYSTEM ASSETS - HIERARCHY

Choose Data Set: ○ Table B-3, Replacement Cost & Asset Value ◉ Table H-2, Performance Scores (raw)

Table H-2: PERFORMANCE SCORES

Level 1	Level 2	Level 3	Level 4	Service Interruptions	Main Breaks	Water Quality	R&R Status	Maintenance Activity	Preventive Maintenance	Age vs. Service Life	Asset Classification	TOTAL SCORE
Distribution System	Central Pipeline	CPL Guadalupe Turnout t	CPL Piping - Sta. 666+09 to 666+47	10	10	10	4	8	10	10	Distribution	9.90
Plant Assets	RWTP Chlorine (OCL) Sy	RWTP OCL Storage Tank	RWTP OCL Storage Tank # 1	10	10	10	10	7	10	0	Plant	7.55
Plant Assets	RWTP Chlorine (OCL) Sy	RWTP OCL Storage Tank	RWTP OCL Transfer Pump # 1	10	10	10	4	10	10	8	Plant	9.60
Plant Assets	RWTP Chlorine (OCL) Sy	RWTP OCL Storage Tank	RWTP OCL Supply Pump # 1	10	10	10	4	7	10	8	Plant	9.15
Plant Assets	RWTP Chlorine (OCL) Sy	RWTP OCL Storage Tank	RWTP OCL Sump Pump # 1 (storage)	10	10	10	4	3	0	0	Plant	5.95
Plant Assets	RWTP Chlorine (OCL) Sy	RWTP OCL Storage Tank	RWTP OCL Storage Tank # 2	10	10	10	10	7	10	0	Plant	7.55
Plant Assets	RWTP Chlorine (OCL) Sy	RWTP OCL Storage Tank	RWTP OCL Transfer Pump # 2	10	10	10	4	10	10	8	Plant	9.60
Plant Assets	RWTP Chlorine (OCL) Sy	RWTP OCL Storage Tank	RWTP OCL Supply Pump # 2	10	10	10	4	7	10	8	Plant	9.15
Plant Assets	RWTP Chlorine (OCL) Sy	RWTP OCL Storage Tank	RWTP OCL Sump Pump # 2 (storage)	10	10	10	4	0	0	0	Plant	5.50
Plant Assets	RWTP Chlorine (OCL) Sy	RWTP OCL Storage Tank	RWTP OCL Storage Tank # 3	10	10	10	10	7	10	0	Plant	7.55
Plant Assets	RWTP Chlorine (OCL) Sy	RWTP OCL Storage Tank	RWTP OCL Transfer Pump # 3	10	10	10	4	0	10	8	Plant	8.10
Plant Assets	RWTP Chlorine (OCL) Sy	RWTP OCL Storage Tank	RWTP OCL Supply Pump # 3	10	10	10	4	7	10	10	Plant	9.55
Plant Assets	RWTP Chlorine (OCL) Sy	RWTP OCL Storage Tank	RWTP OCL Sump Pump # 3 (storage)	10	10	10	4	10	0	0	Plant	7.00
Plant Assets	RWTP Chlorine (OCL) Sy	RWTP OCL Room 224	RWTP OCL Exhaust Fan	10	10	10	4	7	10	8	Plant	9.15
Plant Assets	RWTP Chlorine (OCL) Sy	RWTP OCL Room 224	RWTP OCL Meter Pump # 1	10	10	10	4	7	10	10	Plant	9.55
Plant Assets	RWTP Chlorine (OCL) Sy	RWTP OCL Room 224	RWTP OCL Meter Pump # 2	10	10	10	4	7	10	10	Plant	9.55
Plant Assets	RWTP Chlorine (OCL) Sy	RWTP OCL Room 224	RWTP OCL Meter Pump # 3	10	10	10	4	10	10	8	Plant	9.60
Plant Assets	RWTP Chlorine (OCL) Sy	RWTP OCL Room 224	RWTP OCL Meter Pump # 4	10	10	10	4	7	10	10	Plant	9.55
Plant Assets	RWTP Chlorine (OCL) Sy	RWTP OCL Room 224	RWTP OCL Meter Pump # 5	10	10	10	4	7	10	8	Plant	9.15
Plant Assets	PWTP Chlorine (OCL) Sy	PWTP OCL Room	PWTP OCL Exhaust Fan	10	10	10	4	8	0	8	Plant	8.30
Plant Assets	PWTP Chlorine (OCL) Sy	PWTP OCL Room	PWTP OCL Meter Pump # 1	10	10	10	4	10	10	8	Plant	9.60
Plant Assets	PWTP Chlorine (OCL) Sy	PWTP OCL Room	PWTP OCL Meter Pump # 2	10	10	10	4	7	10	8	Plant	9.15
Plant Assets	PWTP Chlorine (OCL) Sy	PWTP OCL Room	PWTP OCL Meter Pump # 3	10	10	10	4	10	10	8	Plant	9.60
Plant Assets	PWTP Chlorine (OCL) Sy	PWTP OCL Room	PWTP OCL Meter Pump # 4	10	10	10	4	10	10	8	Plant	9.60
Plant Assets	PWTP Chlorine (OCL) Sy	PWTP OCL Room	PWTP OCL Meter Pump # 5	10	10	10	4	10	10	10	Plant	10.00

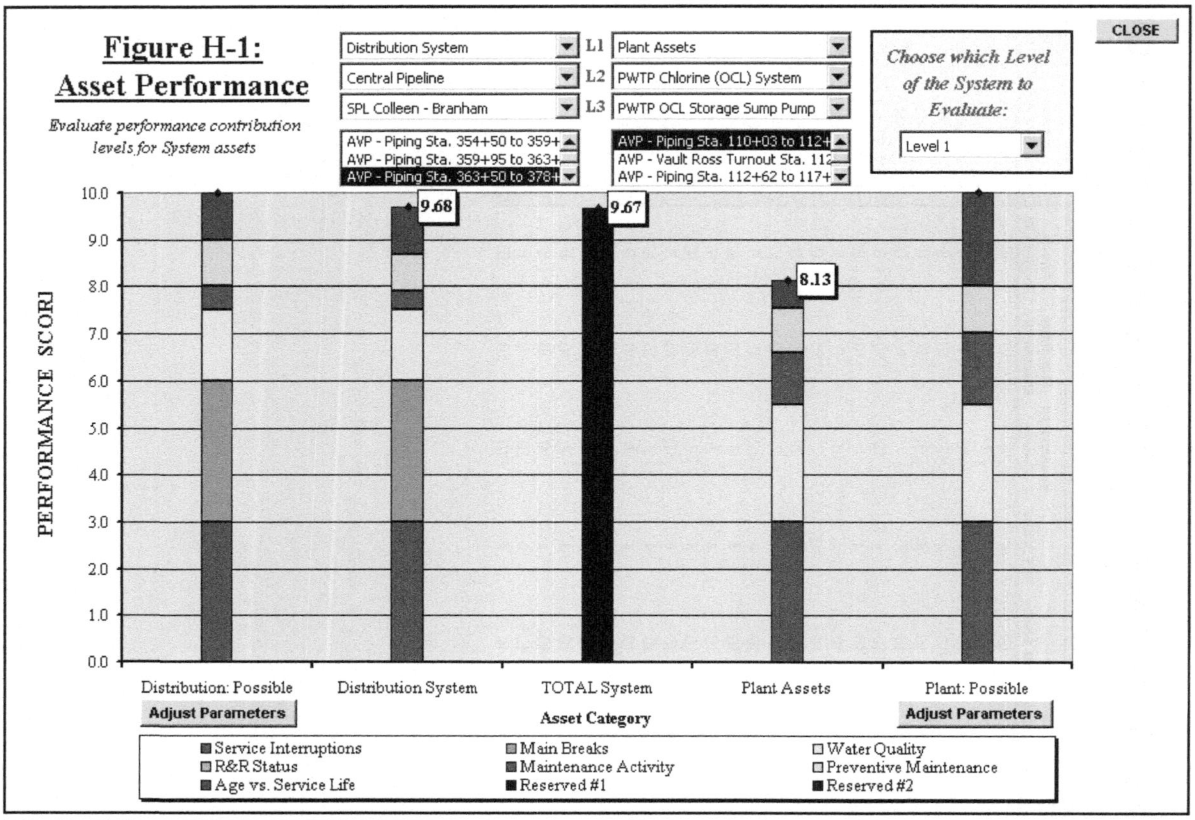

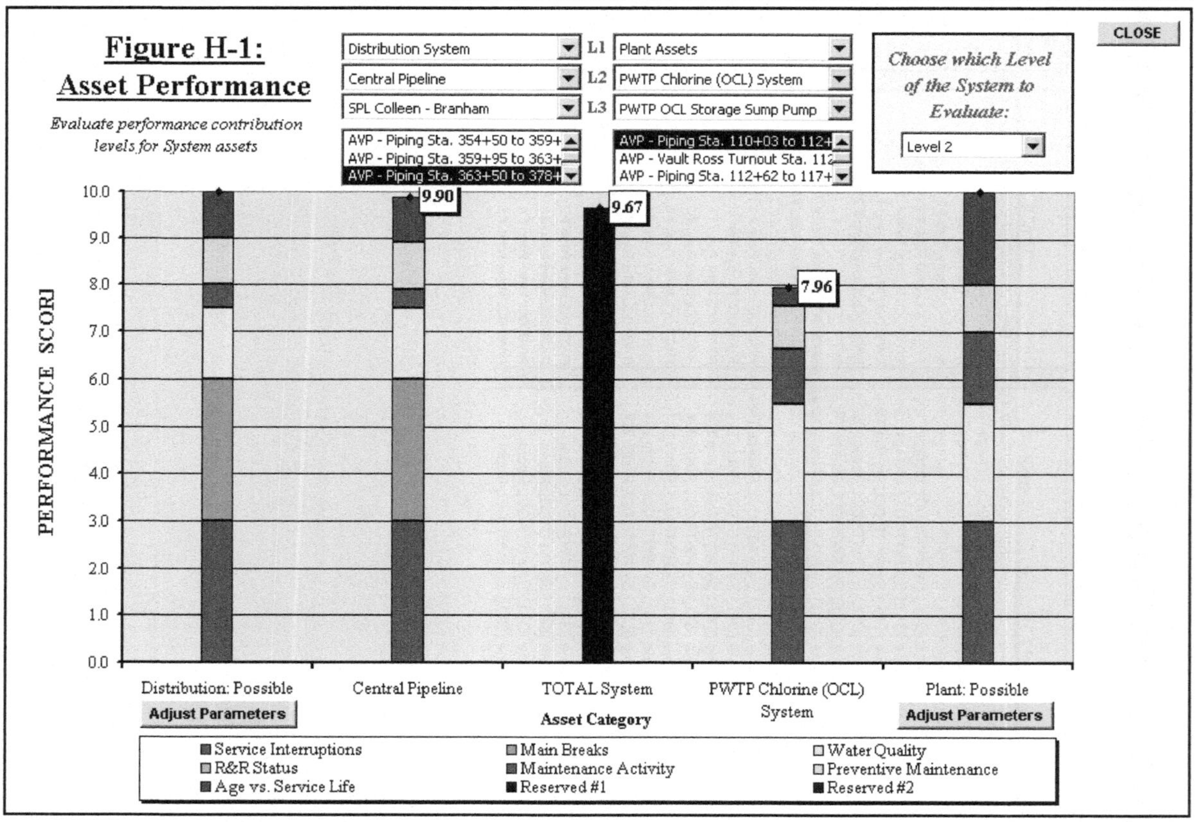

Figure H-1 Asset performance (text on pages 38-39 details cases used in this illustration)

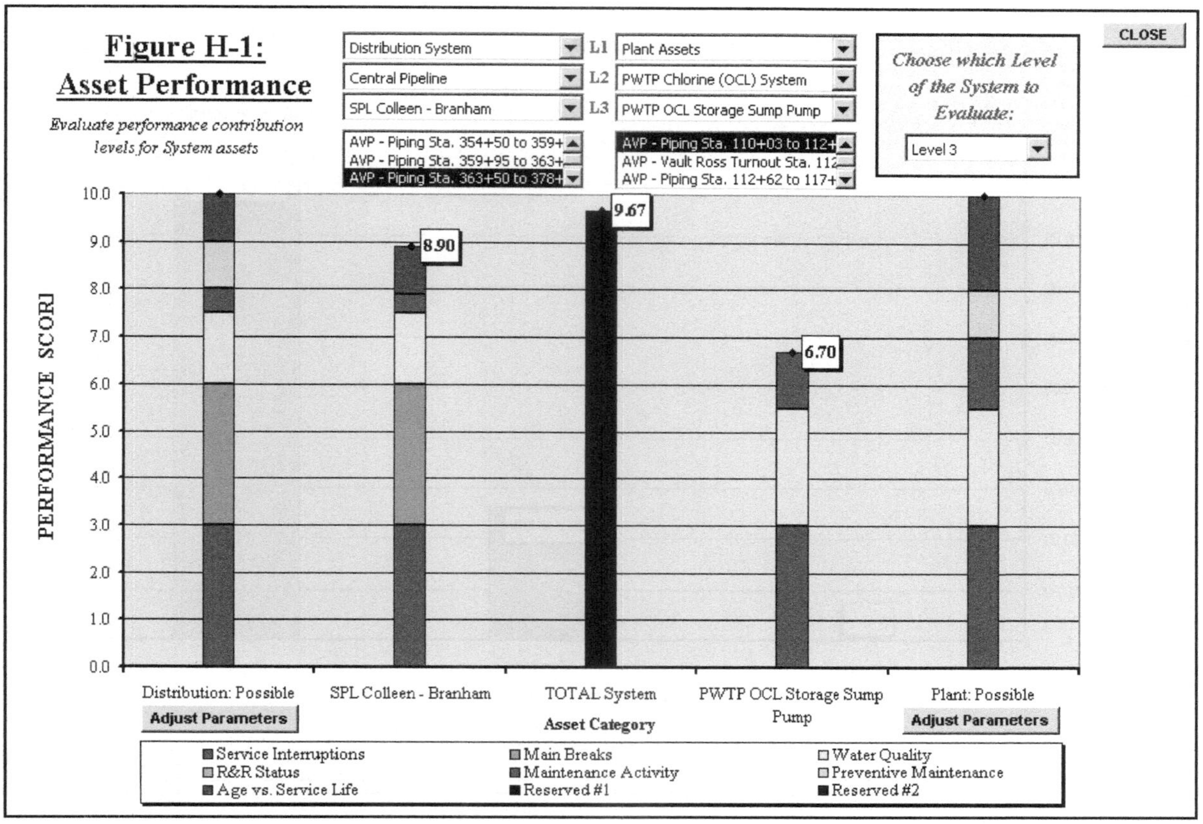

Figure H-1 (Continued)

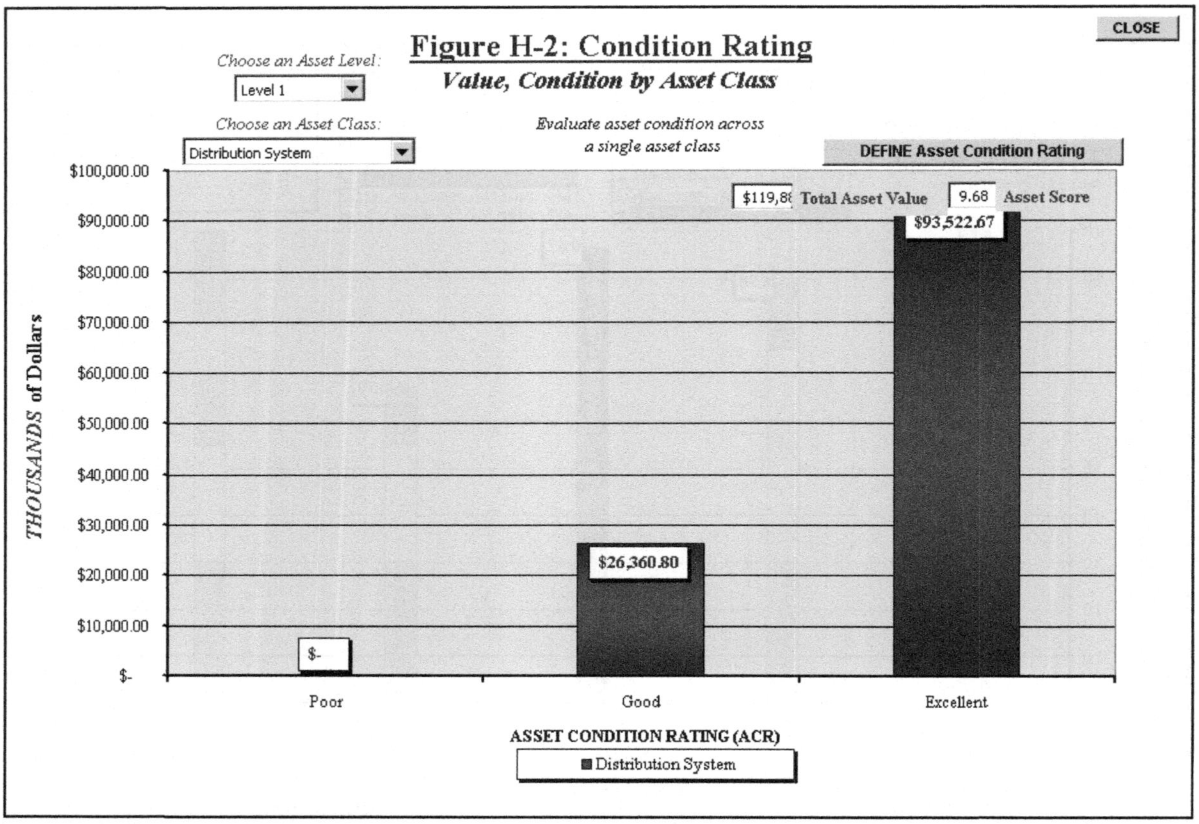

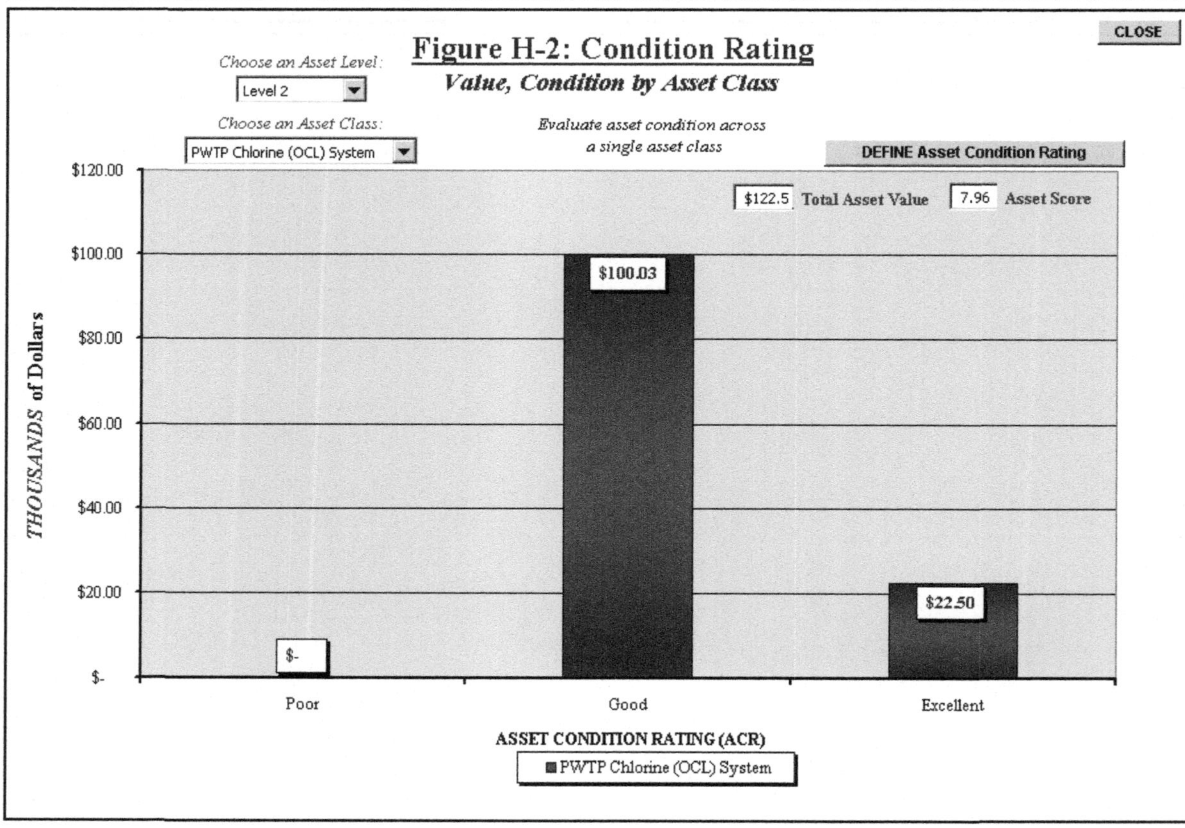

Figure H-2 Condition rating (text on page 39 details three cases used in this illustration)

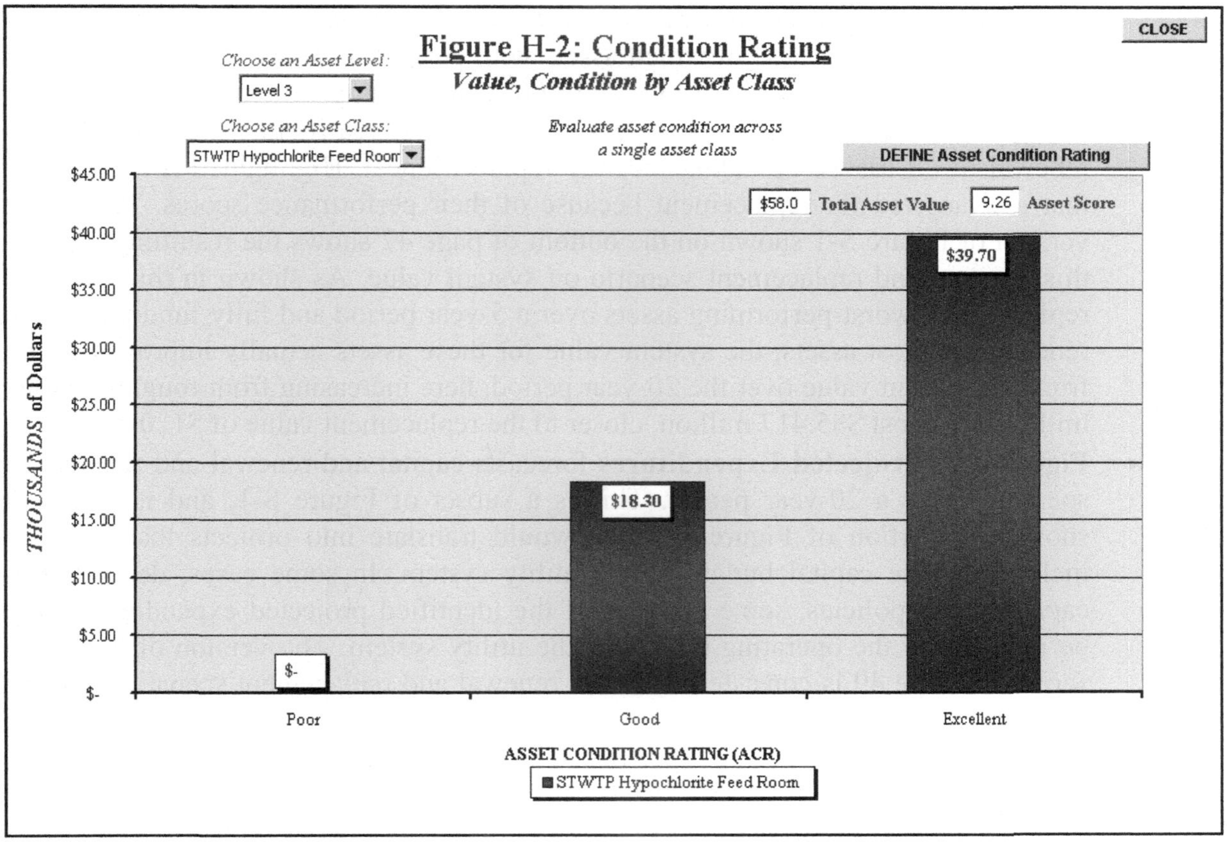

Figure H-2 (Continued)

STRATEGIC OPTION

The several tables and figures developed for the strategic option build upon the weighted performance measure scoring that was developed for the high-end option by adding a time dimension and considering the effect of targeted R&R or full replacement strategies on the estimated value of the system. The tables and figures are listed below; examples follow:

- **Figure S-1: Net System Value** (no replacement or renewal investment) projects net system value over a 20-year period on the basis of various levels of reinvestment and R&R spending proposed in Table S-1. In the example illustrated on the top of page 47 no additional replacement or ongoing renewal work is assumed. As shown in the figure, this results in a reduction in the value of the portion of SCVWD's assets included in this analysis from roughly $84 million to approximately $59 million over the 20-year period. The target (replacement) value and estimated current system value are plotted against the scale shown along the left border, whereas the projected replacement and R&R expenditures are plotted using the scale shown along the right.
- **Table S-1: Platform for Asset Replacement Planning** reports total performance score, replacement cost, target R&R, and remaining value for individual assets of the system. The table highlights underperforming assets and estimates various levels of capital expenditures and R&R spending based on renewal and replacement decisions.

For example, for the portion of SCVWD's asset included in this study (p. 48), a scenario has been created in which assets with performance scores lower than eight are replaced over a 5-year period, and 100% of ongoing renewal expenditures are budgeted for assets with scores less than 10 (essentially fully funding ongoing renewal of all assets, including recently replaced ones). Highlighted rows are those that are targeted for replacement because of their performance scores. The revised version of Figure S-1 shown on the bottom of page 47 shows the resulting impact of this renewal and replacement scenario on system value. As shown in this figure, by replacing the worst-performing assets over a 5-year period and fully funding ongoing renewal for most assets, the system value for these assets actually improves slightly from the current value over the 20-year period, here increasing from roughly \$84.044 million to almost \$85.417 million, closer to the replacement value of \$120 million.

- **Figure S-2: Projected Expenditures** forecasts capital and renewal and replacement spending over a 20-year period. This is a subset of Figure S-1, and it essentially shows the portion of Figure S-1 that would translate into projects that would be included in the capital budget of the utility system. In some cases, depending on capitalization policies, some portions of the identified projected expenditures might be included in the operating budget of the utility system. The version of Figure S-2 shown on page 49 is consistent with the renewal and replacement scenario described above for Table S-1.

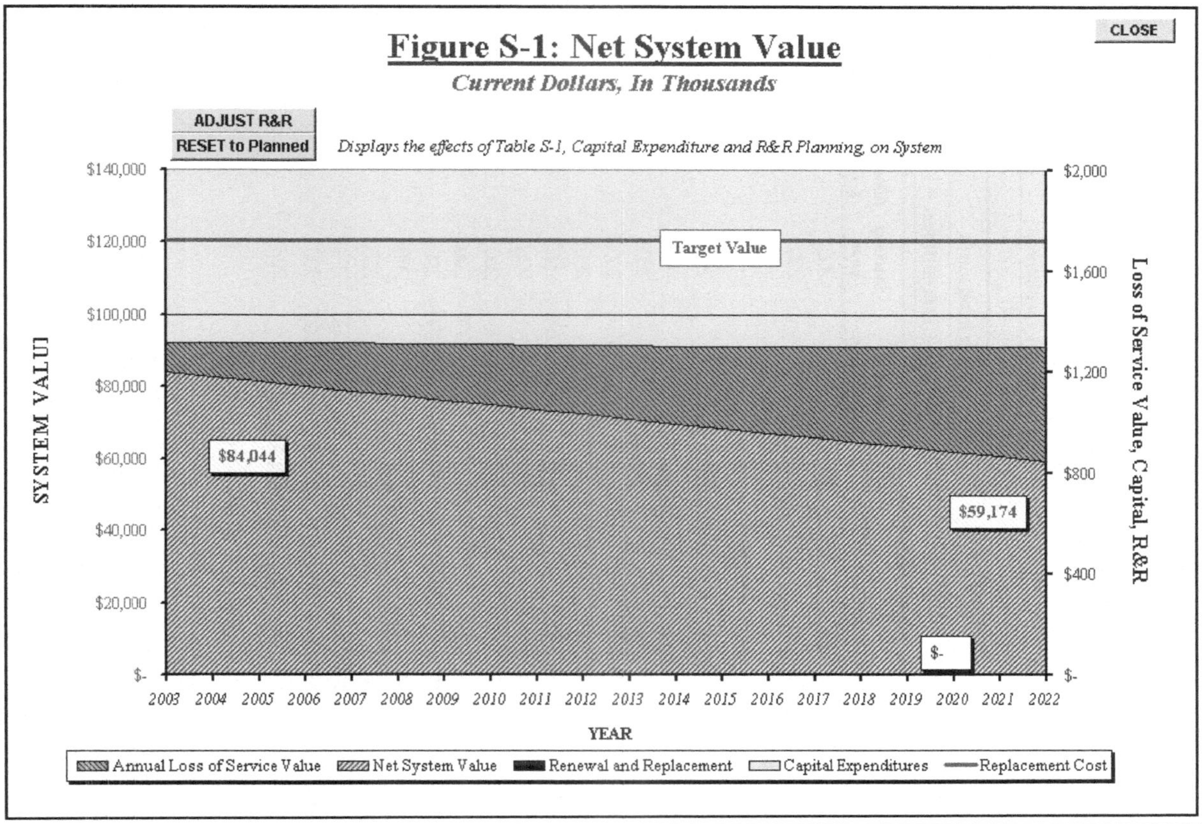

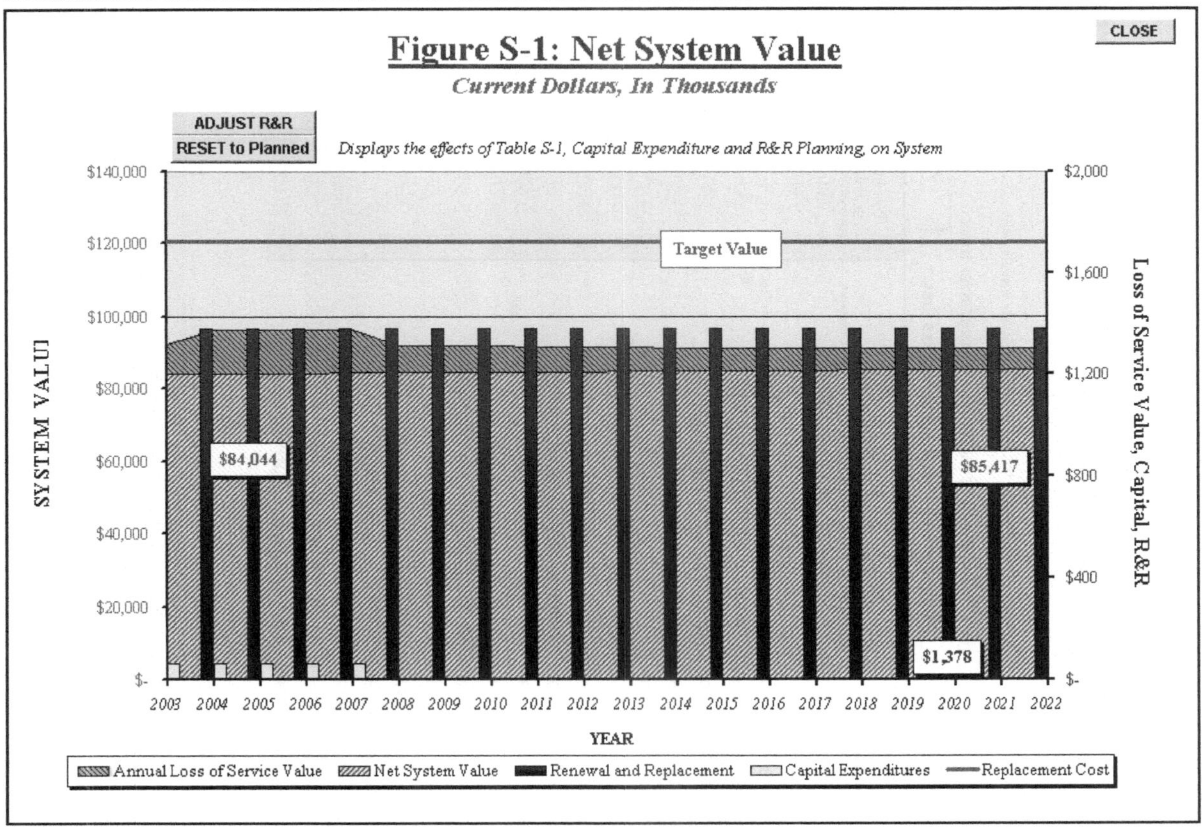

Figure S-1 Net system value (text on pages 45-46 details two cases used in this illustration)

Table S-1
Platform for asset replacement planning

Asset Management Planning and Reporting Options (AMPRO)
Santa Clara Valley Water District

Table S-1: PLATFORM for ASSET REPLACEMENT PLANNING

SCHEDULE | 100 | Percent of Target R&R for assets with performance scores lower than: | 10

SCHEDULE | 50 | Percent of Target R&R for assets with remaining value less than: | 5 | percent

HIGHLIGHT | REPLACE | All assets with performance scores lower than: | 8 | the next: | 5 years

HIGHLIGHT | REPLACE | All assets with remaining value less than: | 5 | ent over the next: | 5 years

Total Capital Expenditures	Annual R&R
$309,180	$1,377,987

CLEAR HIGHLIGHTS | RESET ALL EXPENDITURES | CHART S-1: SYSTEM VALUE

Level 3	Level 4	TOTAL SCORE	Replacement Cost	TARGET R&R	REMAINING VALUE	Capital Expenditure	Starting Year	Allocated X Years	PLANNED R&R
CPL Guadalupe Turnout to Vas	CPL Vault - Page Turnout Sta. 646+00	8.85	$ 11,395	$ 380	$ -	$ -	2003	1	$ 380
CPL Guadalupe Turnout to Vas	CPL Piping - Sta. 646+00 to 656+58.50	9.90	$ 634,800	$ 6,900	$ 372,600	$ -	2003	1	$ 6,900
CPL Guadalupe Turnout to Vas	CPL Vault - Kirk Ditch Turnout & Guard Valve Sta. 656	8.85	$ 7,600	$ 253	$ -	$ -	2003	1	$ 253
CPL Guadalupe Turnout to Vas	CPL Piping - Sta. 656+58.50 to 658+30	9.90	$ 103,200	$ 1,122	$ 60,574	$ -	2003	1	$ 1,122
CPL Guadalupe Turnout to Vas	CPL Piping - Sta. 658+30 to 659+75	9.90	$ 87,000	$ 946	$ 51,065	$ -	2003	1	$ 946
CPL Guadalupe Turnout to Vas	CPL Piping - Sta. 659+75 to 665+22	9.90	$ 328,200	$ 3,567	$ 192,639	$ -	2003	1	$ 3,567
CPL Guadalupe Turnout to Vas	CPL Piping - Sta. 665+22 to 665+82	9.90	$ 36,000	$ 391	$ 21,130	$ -	2003	1	$ 391
CPL Guadalupe Turnout to Vas	CPL Piping - Sta. 665+82 to 666+09	9.90	$ 16,200	$ 176	$ 9,509	$ -	2003	1	$ 176
CPL Guadalupe Turnout to Vas	CPL Piping - Sta. 666+09 to 666+47	9.90	$ 22,800	$ 248	$ 13,383	$ -	2003	1	$ 248
RWTP OCL Storage Tank # 1	RWTP OCL Storage Tank # 1	7.55	$ 42,290	$ 8,458	$ -	$ 42,290	2003	5	$ 8,458
RWTP OCL Storage Tank # 1	RWTP OCL Transfer Pump # 1	9.60	$ 2,100	$ 140	$ 1,120	$ -	2003	1	$ 140
RWTP OCL Storage Tank # 1	RWTP OCL Supply Pump # 1	9.15	$ 5,500	$ 367	$ 2,933	$ -	2003	1	$ 367
RWTP OCL Storage Tank # 1	RWTP OCL Sump Pump # 1 (storage)	5.95	$ 1,000	$ 200	$ -	$ 1,000	2003	5	$ 200
RWTP OCL Storage Tank # 2	RWTP OCL Storage Tank # 2	7.55	$ 42,290	$ 8,458	$ -	$ 42,290	2003	5	$ 8,458
RWTP OCL Storage Tank # 2	RWTP OCL Transfer Pump # 2	9.60	$ 2,100	$ 140	$ 1,120	$ -	2003	1	$ 140
RWTP OCL Storage Tank # 2	RWTP OCL Supply Pump # 2	9.15	$ 5,500	$ 367	$ 2,933	$ -	2003	1	$ 367
RWTP OCL Storage Tank # 2	RWTP OCL Sump Pump # 2 (storage)	5.50	$ 1,000	$ 200	$ -	$ 1,000	2003	5	$ 200
RWTP OCL Storage Tank # 3	RWTP OCL Storage Tank # 3	7.55	$ 42,290	$ 8,458	$ -	$ 42,290	2003	5	$ 8,458
RWTP OCL Storage Tank # 3	RWTP OCL Transfer Pump # 3	8.10	$ 2,100	$ 140	$ 1,120	$ -	2003	1	$ 140

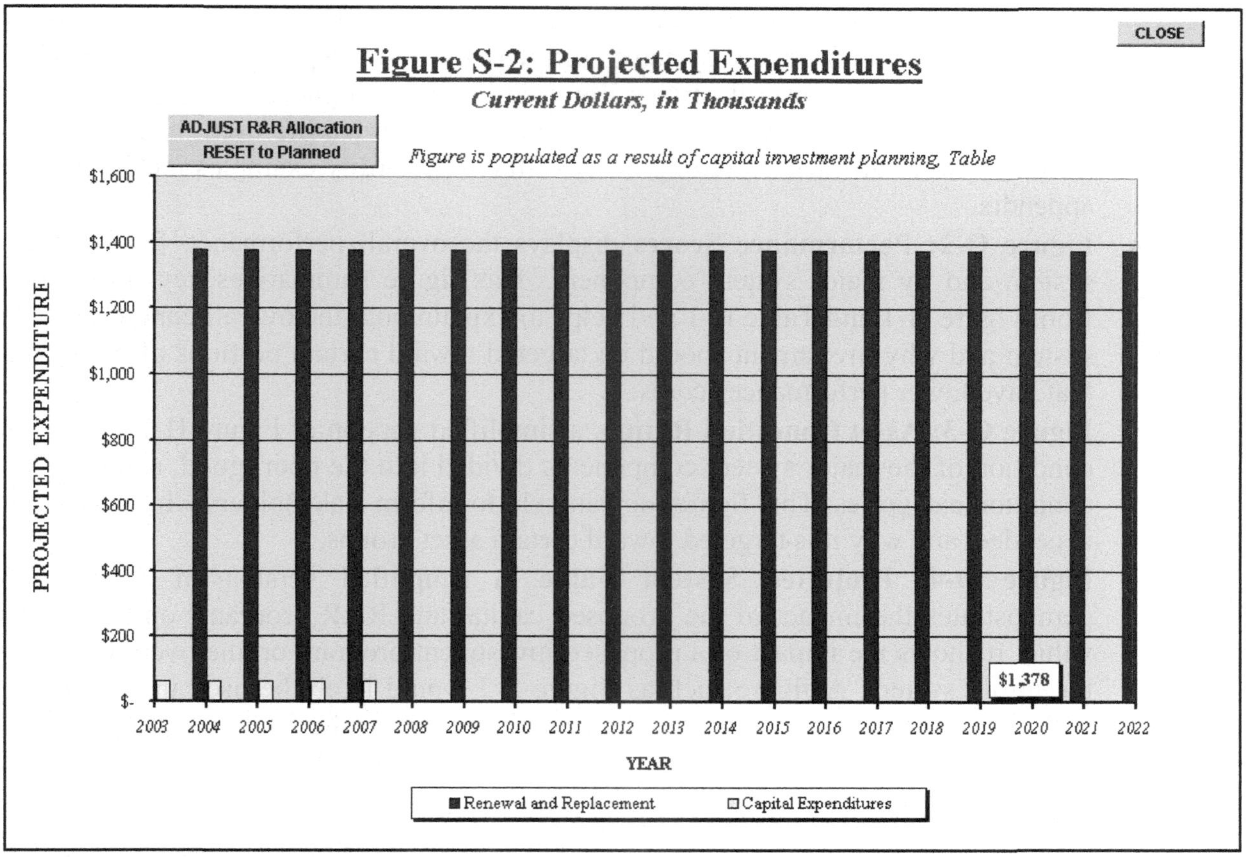

Figure S-2: Projected Expenditures

OUTREACH GRAPHICS

During the project workshop sessions related to audiences and the kinds of information and presentations that are useful in communicating asset management needs, it became clear that most of the figures and tables presented in the preceding portions of this chapter could be useful to utility staff members in crafting an asset management strategy. It also became clear that more simplified ways to portray the results of these technical analyses were needed in order to communicate effectively with most of the external stakeholder groups whose support would be needed to secure the funding for identified renewal and replacement programs.

The four outreach figures described below were developed to illustrate the types of graphics that could be developed to communicate the results of the asset management options studied during the course of this project. These four sample outreach figures do *not* represent the full universe of ways that information about the technical options could be portrayed to stakeholders, but the Project Team determined that they represent reasonable examples that any utility could start from. When implementing the types of asset management options studied in this project, a utility could then modify and add to this initial set of outreach graphics in order to meet the communication needs of its specific stakeholder groups.

- **Figure O-1: Service Life Used** shows the service life used by major system component and by system. This figure incorporates key summary information from

Figure B-1 and Table B-1 and could be used to help explain the need for investment if a significant portion of key systems' services lives are depleted. For the SCVWD example (p. 51), the overall service life used is about 30 percent. The SCVWD is a relatively new system, and as a result it shows less service life used than is the case for many water utilities, including many of the other utility examples included in the appendix.

- **Figure O-2: Performance Scores** displays the overall performance for the entire system and by major system component. This figure summarizes key information from Figure H-1 and Table H-1 and helps to explain both the overall condition of the system and why investment should be targeted toward certain portions of the system that have lower performance scores.
- **Figure O-3: Asset Condition Rating**, a simplified version of Figure H-2, shows the condition of the major system components divided into the poor, good, and excellent condition categories. This figure too can help to inform stakeholders why investment is needed, and why it is targeted toward certain asset groups.
- **Figure O-4: Projected System Value**, a simplified version of Figure S-1, demonstrates the impact of the proposed capital and R&R programs on net system value. It shows the impact of a proposed investment program on the overall value of the utility system. A figure such as Figure S-2 would likely be used in conjunction with this figure to identify the funding requirements for the recommended program. The proposed program shown in the version of Figure O-4 is based on the renewal and replacement described in Table S-1.

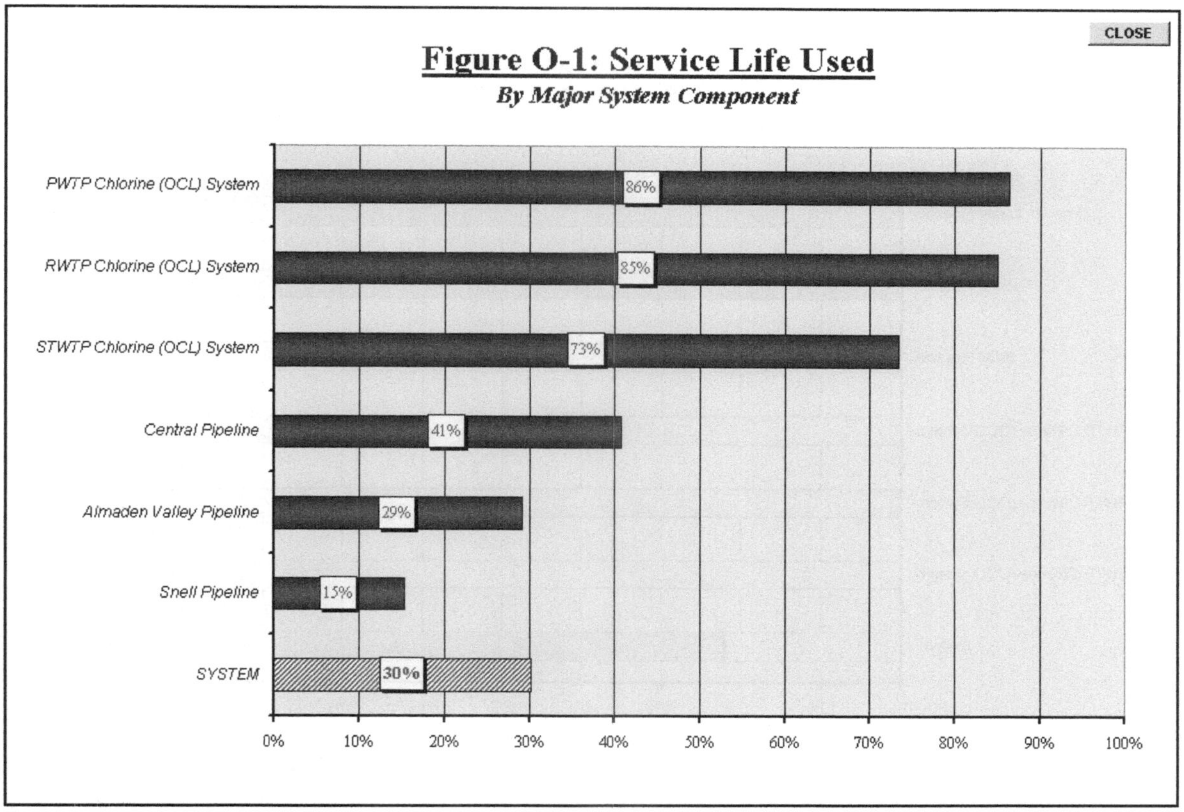

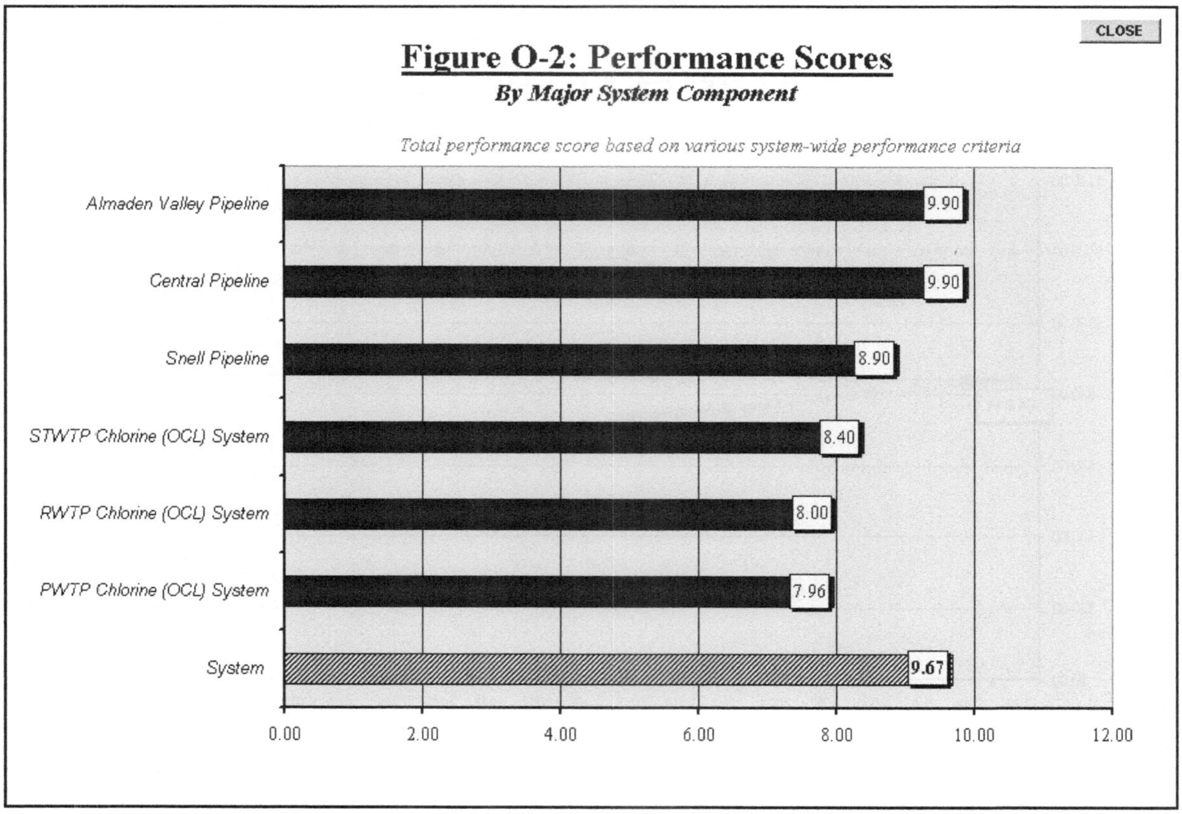

Figure O-1 Service life used

Figure O-2 Performance scores

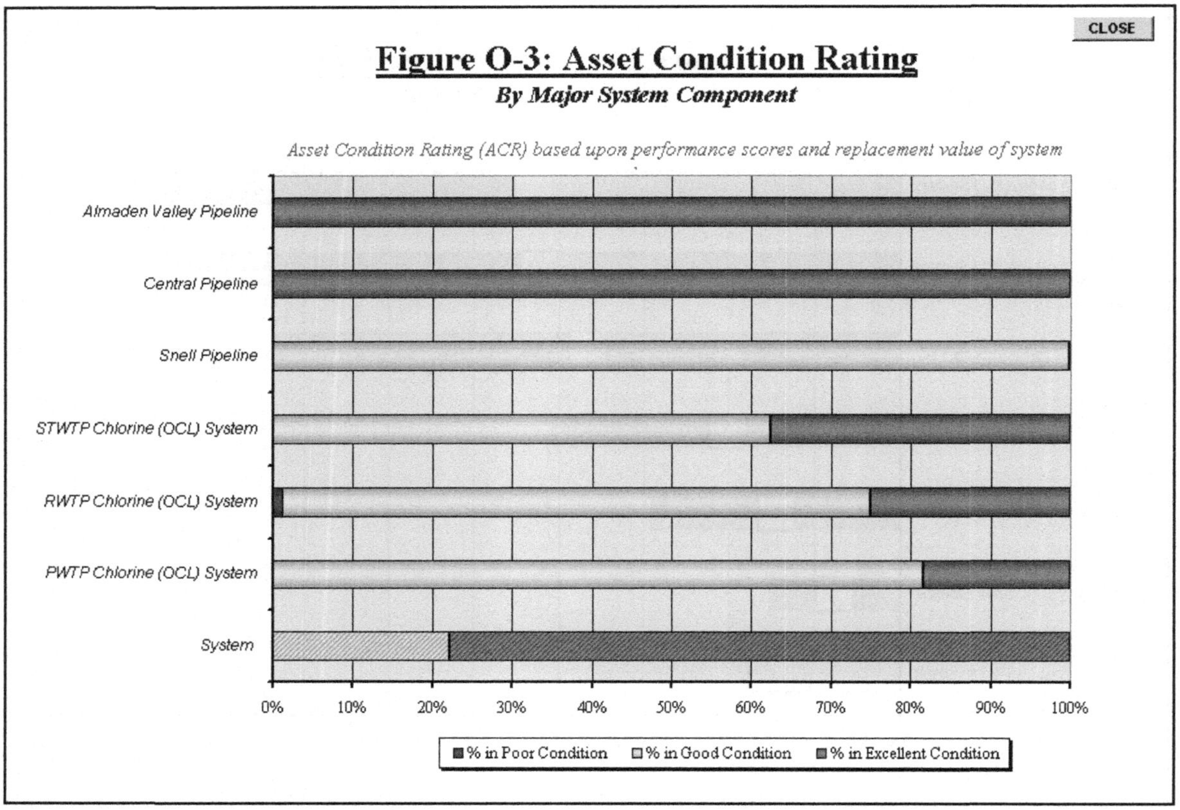

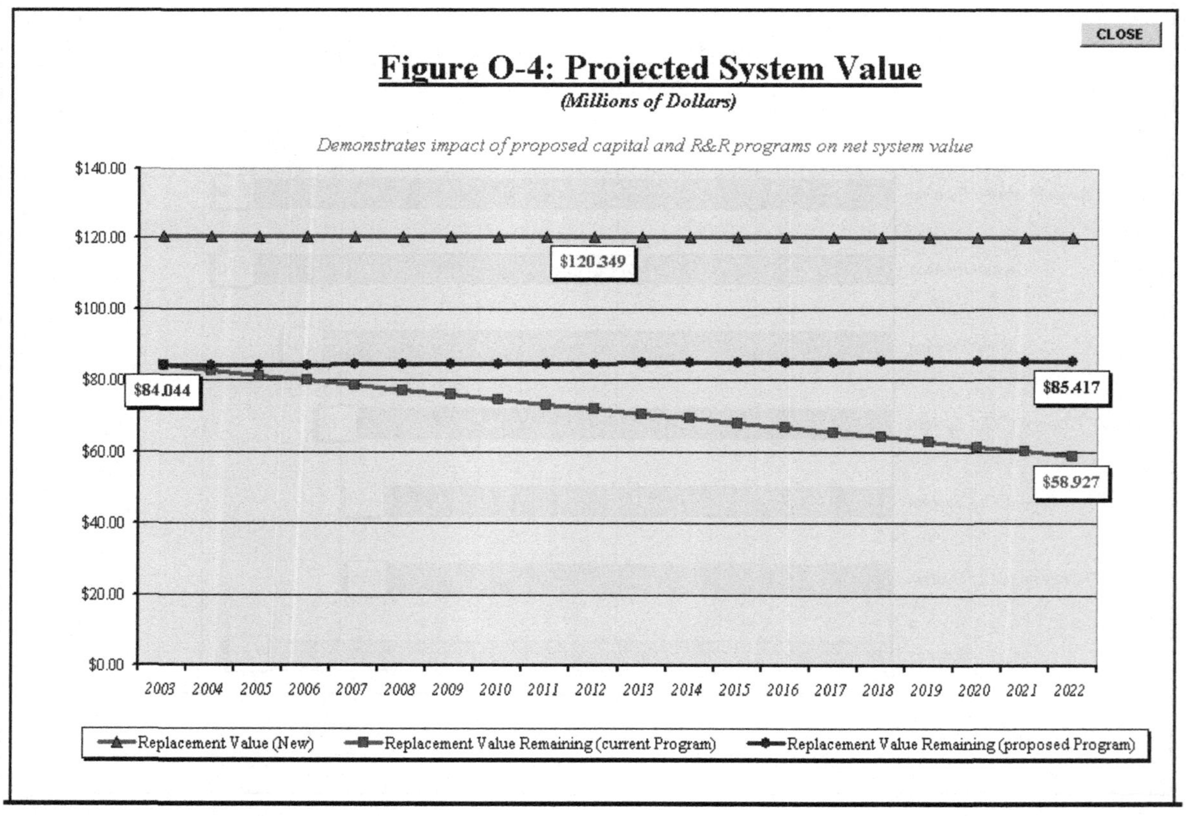

Figure O-3 Asset condition rating

Figure O-4 Projected system value

CHAPTER 4
DEFINING HIGH PERFORMANCE: MEASURING SUCCESS

Highly competitive organizations are able to measure their performance on several levels (AMSA, 2002). The need to measure performance is increasingly important in a competitive environment. Every manager has heard "What gets measured gets done" and "If you don't measure it, you can't manage it."

Success is best defined as an organization attaining its business mission and strategies. An organization's mission is an expression of its purpose. Missions are implemented by well-planned strategies, which are developed to ensure that the mission is accomplished under potentially changing circumstances. The strategy is implemented in the form of initiatives or programs. Each business unit helps support these programs at different levels, depending on the program. Business units make decisions and in doing so implement a strategy. In this way, strategy is implemented in alignment with the organization's mission, as illustrated in Figure 4.1. While this seems very straightforward, very few organizations work in this way.

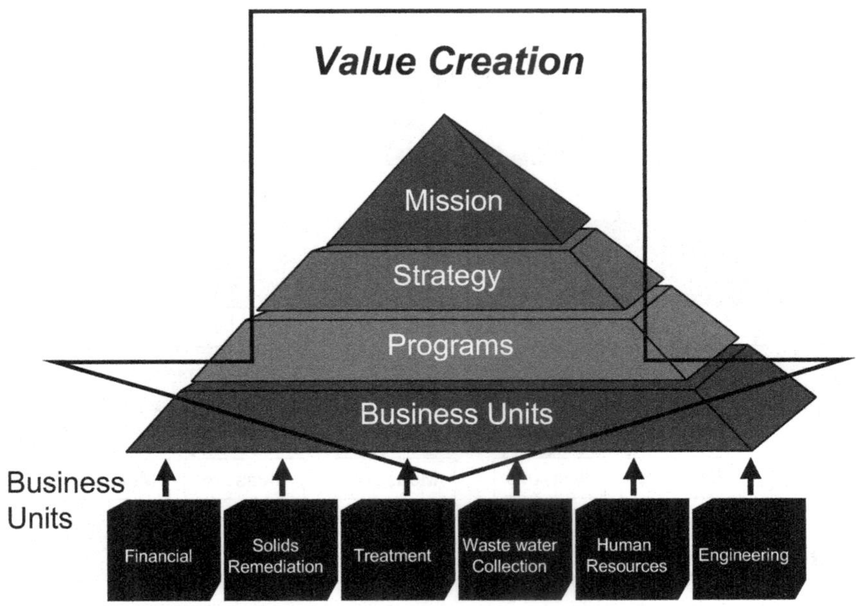

Figure 4.1 Individual business units within a utility support overall mission and strategy

Having a clear mission helps an organization create a mindset that encourages the development of beneficial alternatives and decisions. The organization's mission is explained in the form of *goals*, which are broadly stated results that help to ensure that the organization's mission is attained. They are not directly measurable. Table 4.1 illustrates the goals frequently found in water utility mission statements. *Objectives* help explain the components of the goals and what is expected of them. *Performance measures* link business unit actions to business objectives, goals, and the mission. Both quantitative and qualitative measures explain the achievements made toward an objective or goal. Figure 4.2 illustrates the relationship between goals, objectives, and measures. Some objectives may have only one measure.

Table 4.1
Common goals and objectives in utility mission statements

Goal	Objectives
Satisfied customers	Satisfy customers
	Operate facilities in a superior manner
	Maintain infrastructure
	Complete projects on time and budget
	Evaluate state of the art-technology
	Maintain good planning for economic development
Employee-friendly work environment	Ensure a safe and healthy work environment
	Collaborate in problem solving
	Encourage employee development
Stable and sustaining rate structure	Establish financial accountability
	Maintain a sound financial program
Compliance	Comply with permits and regulations
Community stewardship	Healthy watershed
	Encourage community outreach
	Support economic development

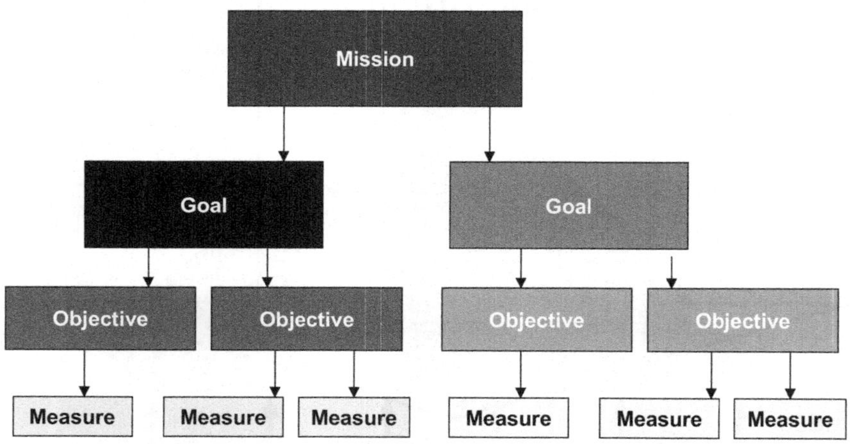

Figure 4.2 Missions drive goals, objectives, and performance

BENEFITS OF A PERFORMANCE MEASUREMENT SYSTEM

There are numerous benefits of properly measuring performance, including:

- Alignment and focus of staff with organizational mission and goals and senior managers' priorities
- Increased accountability to customers and the public
- Improved service effectiveness, quality, and efficiency
- Targets and incentive structures to motivate staff
- Feedback on individual team or department performance
- Balanced financial and nonfinancial measures in decision-making

- Better decisions
- Trust and credibility among stakeholders

When organizations don't know what to measure, they often try to measure too much. A major challenge for organizations is determining which of the hundreds—if not thousands—of financial and nonfinancial measures to track. Much time is wasted collecting data that cannot be used in measuring important performance issues or in making decisions. The result is often that measures are not used to improve the organization or make better choices.

A proactive and competitive organization establishes a small number of relevant measures that are balanced across all their core services and linked directly to strategies and mission.

A major benefit of developing performance measures is improved decision making that aligns the organization's decisions to implementing the organization's mission or new strategies. The more decisions a performance measure influences, the more likely it is that the level of performance will be realized.

Performance measures role in decision making can range from providing baselines and targets to benefit descriptions and scales used in making benefit-cost trade-offs. Mission and performance measures help align and drive decisions on strategy, operations, risk management, and continuous improvement (Figure 4.3).

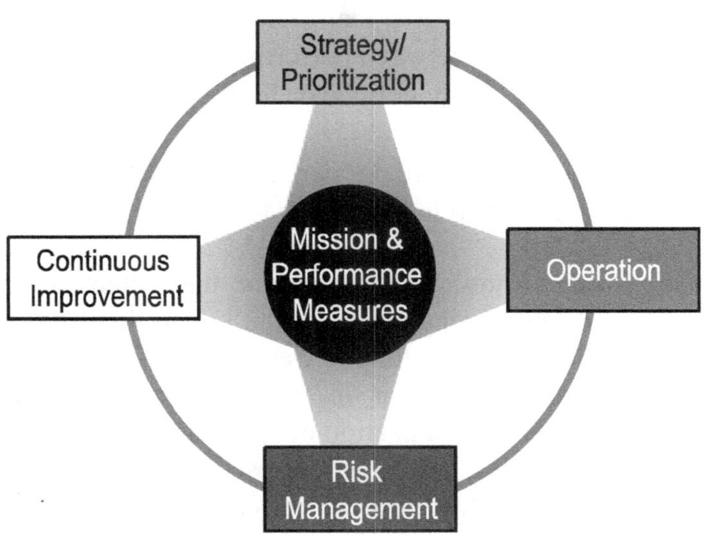

Figure 4.3 Role of performance measures

Five key decision processes that performance measures can significantly improve are listed in Table 4.2.

Table 4.2
Role of performance measures in decision making

Decision	Role	Business issues
Operations decisions	Baselines, targets	Compliance, reliability, continuous improvement
Strategy selection	Excellent alternatives	Direct new strategy
Risk assessment	Define severity or consequence of risk	Protect mission, manage risk, save money
Benefit/cost	Define benefits and scales	Alternatives selection, prioritization, capital planning
Systems thinking	Baselines and targets	Limiting factors in system performance

Many utility organizations believe they have some form of performance measurement system already in place. Yet, these systems rarely provide the results expected of them. Many organizations measure efficiency and not necessarily what customers value about their services, or which program is supported by the service delivery.

This may result in an efficient service that may not meet customer needs. Problems found in performance systems include the following:

- Many utilities have an excessive number of measures
- No clear link exists between performance measures and mission and goals
- No clear relationship exists between performance measures and decision making
- Measures are not appropriately defined or focused
- No clear targets
- Measurement and monitoring are not part of many utilities' daily business practice
- Measurement is not linked to individual or team performance

When organizations don't know what to measure, they often measure too much. A major challenge for organizations is determining which of the hundreds, if not thousands, of financial and nonfinancial measures to track. Much time is wasted collecting data that can not be used in measuring important performance issues or in making decisions. The result is often that measures are not used to improve the organization or make better choices.

A proactive and competitive organization establishes a small number of relevant measures that are balanced across all their core services and linked directly to their strategies and their mission. On the basis of the research team's experience in working with water utilities throughout the United States, an appropriate number of performance measures for many utilities is between 20 and 30.

Knowing what to measure is difficult. Measures may change as new issues come up. Some may be "retired" as old problems are solved. A set of performance measures will likely remain stable. These are the ones that are tied to fundamental goals within the organization's mission. The next step is to ensure the level of definition and focus needed for valid and reliable measurement. *Validity* refers to the extent to which a metric succeeds in capturing what it is supposed to capture, while *reliability* refers to the degree to which measurement techniques reveal actual performance changes and do not introduce error of their own. For example, many organizations attempt to use surveys to assess extremely complex performance dimensions such as customer satisfaction and rely only on one or a few questions. The questions may allow respondents only a small number of scale points (e.g., 1, low; 2, medium; 3, high). Many organizations then collapse these answers into crude binary scales (for example, satisfied with a score of 4 or 5 and dissatisfied with a score of 1–3). Although inexpensive to use and easy to understand, such simplistic surveys lack validity and reliability and impair an organization's ability to discern superior performance or predict financial results.

TYPES OF PERFORMANCE MEASURES

Performance measures link the business framework and generally fall into one of three types: effectiveness, efficiency, and quality (Table 4.3).

Table 4.3
Types of performance measures

Performance Measure	
Effectiveness (Adequacy)	Meeting customer needs, achieving customer outcomes, contribution of service value to program goal
Efficiency	Ratio of outputs to inputs (lower cost, faster delivery, smarter)
Quality (Dependability)	Adding value; comparison of service delivery to standard

Source: Adapted from Association of Metropolitan Sewerage Agencies and Association of Metropolitan Water Agencies, *Thinking, Getting, Staying Competitive: A Public Sector Handbook*, 1998.

Programs and business units are the keys to management. Services support programs. Core business processes support services. Resources are allocated to implement core processes, strategy and programs. Effectiveness, quality and efficiency relate to each of these levels.

Performance measures can be selected based on a set of criteria. This ensures a defensible and results-orientated performance measurement program Table 4.4 illustrates criteria that are often relevant to the selection of appropriate performance measure. Another important issue relates to control. There are situations where performance measures are measuring results that people can only partially affect. The less that people can affect the measure, the greater the risk for the irrelevancy of the measure.

Table 4.4
Performance measure selection criteria

Criterion	Definition(s)
Mission driven	Relate directly to a mission element and objective
Comprehensive/balanced	Touch on all key business elements
Linked to value	Cause and effect relationship is understood to ensure usefulness of measure
Relevant	Key stakeholders care about this measure.
Useful	The accuracy of the measure is understood and believed by decision-makers in key decisions
Applicable	Responds to changes at a scale appropriate to the goal or issue
Integrates effects	Integrates effects and exposure over time and space
Valid	Measure the effect that impacts on the organization's goals or success

(continued)

Table 4.4 (Continued)

Criterion	Definition(s)
Reliable	Relevant feature having defined numerical scales and can be quantified simply or qualitative scales that measure nonquantifiable measures in a meaningful manner. The resolution of the scales is appropriate to discriminate meaningful differences. This may also relate to surrogates that are representative of a group of features
Mutually exclusive	No double counting
Reproducible	Reproducible within defined and acceptable limits of data collection over time and space
Understandable	Measures can be transformed into a format that target audiences can understand
Obtainable	Able to gather and expertise to monitor
Cost-effective	Information is available or can be obtained with reasonable cost/effort

THE BALANCED SCORECARD

One useful method of developing internal performance measures is called the "Balanced Scorecard" concept. Devised by Robert Kaplan and David Norton of the Harvard Business School, the balanced scorecard method results in a set of measures intended to give top managers a fast and comprehensive view of the business. It includes financial measures that tell the results of actions already taken and operational measures on customer satisfaction.

Many organizations only measure what is obvious or easy to collect or needs to be collected for regulatory purposes or financial measures that ratepayers ask about. When there are performance measures in each key business process, one is "balancing the scorecard." The notion of a balanced scorecard means that each key business process is being adequately measured for effectiveness, efficiency, and quality to ensure high performance.

The balanced scorecard allows a manager to look at the business from four important perspectives:

1. How do customers see us?
2. What must we excel at?
3. Can we continue to improve and create value?
4. How do we look to our shareholders?

Having a balanced list of performance measures that addresses all aspects of the mission helps to strengthen and reinforce the right actions (Figure 4.4):

> Think of the balanced scorecard as the dials and indicators in an airplane cockpit. For the complex task of navigating and flying an airplane, pilots need detailed information about many aspects of the flight—fuel, air speed, altitude, bearing, destination, and other indicators—to summarize the current predicted environment. Reliance on one instrument can be fatal. Similarly, the complexity

of managing an organization today requires that managers be able to view performance in several areas simultaneously.[1]

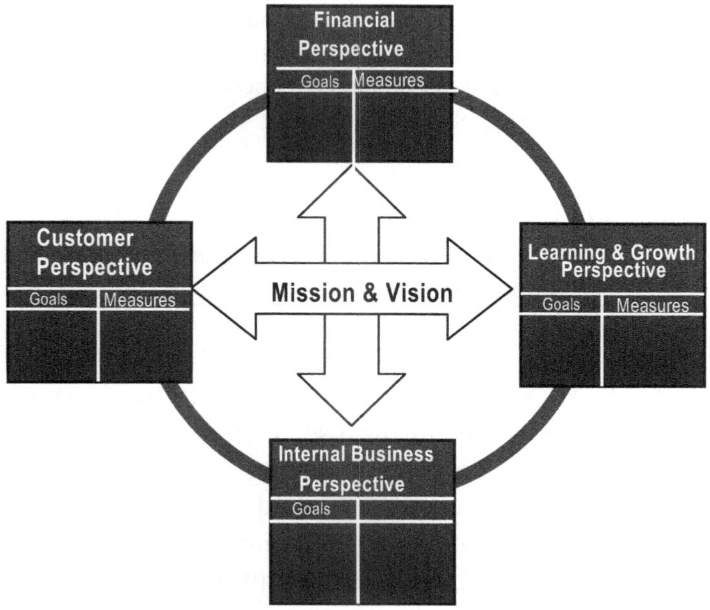

Figure 4.4 Balanced score card emphasizing all aspects of business performance to ensure a successful mission

Finally, measuring what key stakeholders care about and effectively communicating the results engenders credibility and trust both inside and outside the organization. When organizations accomplish pre set performance measures and targets, key external stakeholders develop a level of trust and are proud of their association with the organization. By measuring the business processes that create real value, staff members are proud to be part of a winning strategy and become motivated to change their behavior and work to create value. These measures define performance in competitive organizations. Decisions are only improved by tying them to missions, goals, and performance measures. The more decisions that a performance measure influences, the more likely it is to realize that level of performance.

Table 4.5 shows the primary measures that were used to test asset performance by the participating utilities during the course of this research project. These measures were selected the team from 47 candidate measures (Table 2.1, p. 14). As shown in Table 4.5, the measures represent a balance among the elements of a balanced scorecard (effectiveness, efficiency, and quality). This balance was by design, as the participating utilities and members of the research team worked to achieve a limited, but balanced, set of measures to test.

[1] R. S. Kaplan and D. P. Norton, "Using the Balanced Scorecard as a Strategic Management System." In *Harvard Business Review*, January-February 1996.

Table 4.5
Performance measures studied in this project

Effectiveness	Efficiency	Quality
Number of unplanned service interruptions	Maintenance activity (planned maintenance costs and reactive maintenance costs as a percentage of replacement value)	Violations of internal standards
Number of main breaks per length of pipe (mile)	Preventive maintenance (scheduled vs. completed)	R&R status (assets relined, rebuilt, or protected)
Additional measures studied by some participating utilities:		
C factor	Water lost to leakage as a percentage of total water supplied	Customer complaints (by cause) related to assets

South East Water LLC (SEWL), an Australian utility company that served as an advisor to this project, is a recognized leader in the use of performance indicators to guide strategic decision-making. As illustrated in a corporate planning document contributed for inclusion in the Appendix to this report, SEWL has devised a limited number of key performance measures for use in assessing performance and planning forward. For the system's water assets, these indicators include the number of unplanned interruptions, the average duration of unplanned interruptions, the percentage of customers with more than five unplanned service interruptions.

RECENT INDUSTRY RESEARCH ON PERFORMANCE INDICATORS

Subsequent to the selection of performance measures and development of participating-utility data samples for this project, AwwaRF released a research report that provides an additional framework that may be useful to readers when selecting a balanced group of performance measures. *Selection and Definition of Performance Indicators or Water and Wastewater Utilities* (Crotty 2004), identifies 20 recommended performance indicators that are grouped into five business-process categories used by AWWA's QualServe program (customer relations, business planning and management, organizational development, water operations, and wastewater operations). These categories represent another way to think about providing balance in the assessment of performance. Not all of the indicators identified in the performance-indicators study relate directly to asset management. However, a number of the measures do related directly or indirectly to asset management, such as:

- System renewal/replacement rate, which measures the degree to which a utility is renewing or replacing its infrastructure
- Return on assets, which indicates the financial effectiveness of a utility

The performance-indicators study therefore represents another useful source of information for utilities seeking to identify asset management-related measures appropriate for their systems.

CHAPTER 5
DEVELOPING REALISTIC DATA MANAGEMENT STRATEGIES

Developing and maintaining the data required to support renewal and replacement planning poses a substantial challenge for most water and wastewater utilities. Maintaining quality information regarding assets in any one element (e.g., asset descriptors, asset condition, asset financial history, etc.) has proven to be a challenge for most utilities. Since the asset management options being studied in this project require information across all of these elements and the ability to relate those data using common identifiers, the challenge is substantially greater than that associated with maintaining single-element information.

The data management issues that arose in populating the data samples resulted in a significant number of findings and recommendations that emerged from this project. These findings and recommendations are detailed in Chapter 9 and Chapter 10, respectively, with additional case details provided by some of the participating utilities in the letters of comment that are included in the appendix. This chapter addresses overall approaches to developing realistic data management strategies that can be successful in maximizing value to water utility organizations by providing assistance in decision-making and supporting external reporting requirements. Recommended approaches are suggested in light of the issues that arose in developing data management for this project's data sets.

CHALLENGE IN DEVELOPING SUCCESSFUL DATA MANAGEMENT PROGRAMS

The development of successful integrated data management for an asset register requires consideration of the multiple purposes for which the information will be used, including (1) supporting specific decision processes regarding renewal and replacement strategies, (2) evaluating higher-level scenarios to make intelligent tradeoffs, and (3) supporting fiscal management and financial planing and budgeting processes.

Learning from Past Failures

In order to be successful, data management programs need to be on time, on budget, and meet end-users' expectations when implemented. Some industry statistics help to frame the challenges associated with implementing new data programs:[1]

- The average large information systems project misses its scheduled deadline by 25 to 50 percent
- Only 26 percent of information systems projects are delivered on-time
- 70 percent of all information systems fail to meet one or more of users' requirements
- 25 percent of all information systems projects are cancelled before being completed

---

[1]Standish Group. *Charting the Seas of Information Technology*. Dennis, Mass.: The Standish Group, 1994.

A survey of more than 8,000 organizations identified some of the key hurdles that need to be overcome to be successful:[2]

- Incomplete requirements (13%)
- Lack of user involvement (12%)
- Lack of resources (11%)
- Unrealistic expectations (10%)
- Lack of executive support (9%)
- Changing requirements (9%)
- Lack of planning (8%)

Integration Across Systems Increases Challenges

There are substantial challenges in developing and maintaining current data for a municipal utility system for even one function or area of information, such as water quality parameters, system outages, and maintenance expenditures. As detailed throughout this report, even the basic option for asset management requires integration of information from multiple sources (asset descriptions and age information, renewal and replacement costs, useful lives, installation dates, etc.). The data requirements for the three options addressed in the study can be summarized as follows:

- **Basic** (service life)—Create an asset register with service life, installed cost or replacement cost estimate, material, age, and other descriptive information to allow estimates of remaining value at a point in time or over time.
- **High-End** (weighted performance measures)—Requires information for basic option plus asset performance for selected indicators and weighting of the indicators, to allow calculation of prioritized performance scores and condition ratings.
- **Strategic** (weighted performance measures plus time dimension)—Requires the same data as the high-end option, plus additional policy and condition analysis inputs to determine the pace at which R&R will be implemented, based on condition ratings, asset "depreciation," and other factors. Requires additional integration with overall strategic policy planning for the utility to determine the timeframe over which improvement or reduced declines in the value of the asset base will be realized.

A fully integrated asset management data platform, therefore, would require a central data repository that stores information that may currently reside in seven or more locations, such as the CMMS, GIS, and other sources illustrated in Figure 5.1.

---

[2]McConnell, S., and Construx Software. *Software Project Survival Guide: How To Be Sure Your First Important Project Isn't Your Last*. Edited by D. Clark and V. Thulman. Redmond, Washington: Microsoft Press, 1998.

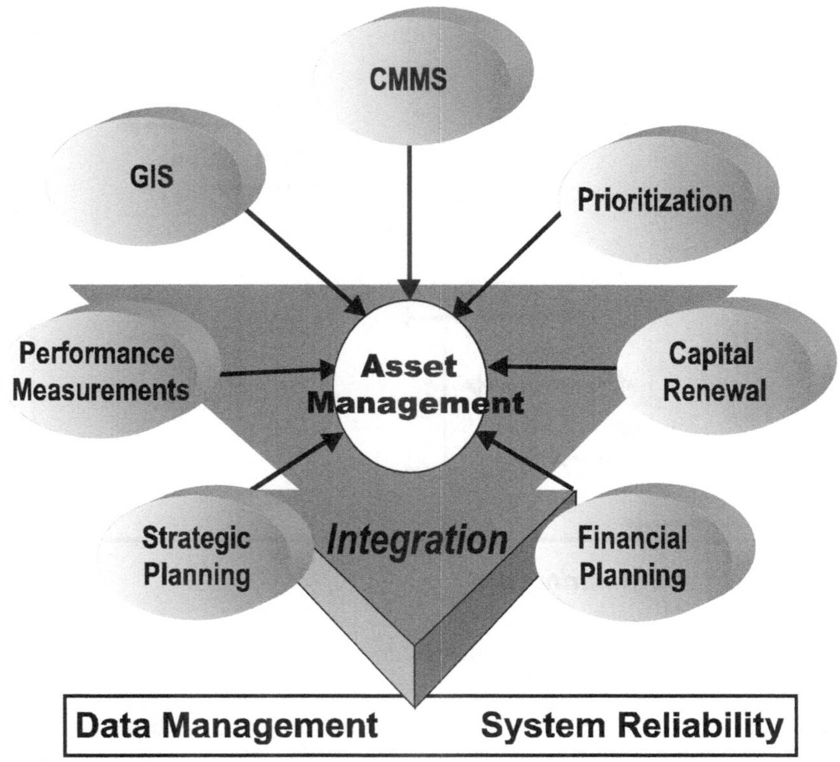

Figure 5.1 A fully integrated asset management data platform

Based on both budget restrictions and staff availability, it is clear that most water utilities cannot immediately develop for all of their assets a fully integrated data platform that covers the full range of descriptive, financial, and performance attributes required to support the asset management options evaluated in this study. The remainder of this chapter and the data management–related discussions in the findings and recommendations chapters identify several pathways that utilities can use in light of this context—for example, developing a fully integrated data platform first for only new assets added to the system or starting initially by developing a fully developed inventory and/or CMMS system. The latter approach follows what can be described as the initial steps in an incremental progression toward a fully integrated asset management data platform, as illustrated in Figure 5.2.

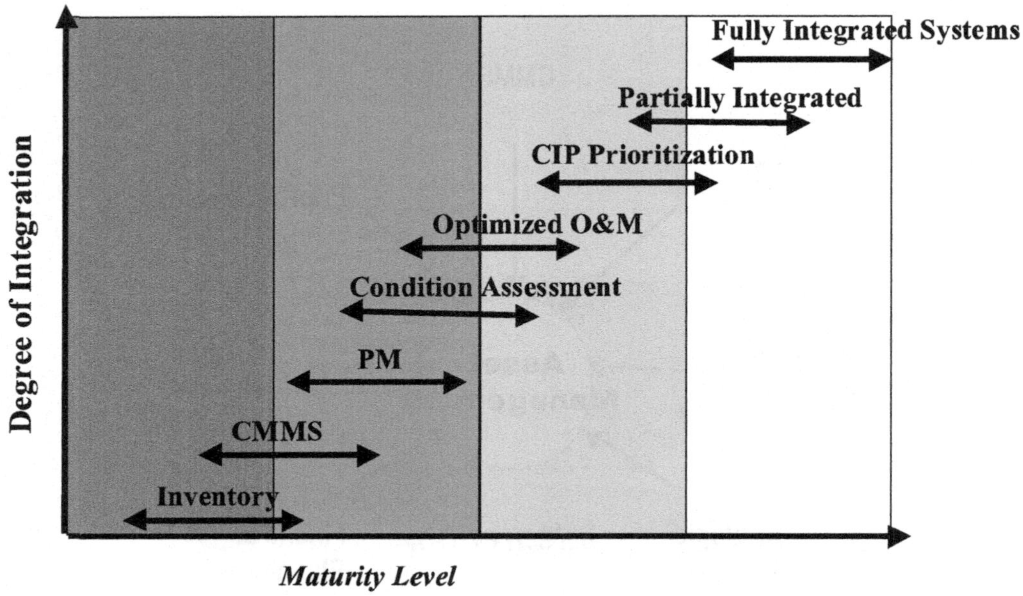

Figure 5.2 Asset management maturity model

This figure is organized to show the progression in the development of information, from basic inventories and CMMS systems up through fully integrated data systems. All of the required information is stored in a central location, and the decision-support programming capabilities required to conduct the analyses illustrated for the strategy option in this study, and more are also available within the same data and programming environment.

Not all water utilities need to develop fully integrated data systems to support their asset management planning objectives. For example, utilities that decide that the basic option implemented at a very high level of asset organization is sufficient to meet management planning and governing board/stakeholder needs may find that a combination of several databases or other models (e.g. inventories, CMMS databases, GIS databases) that can electronically "share" information through transfer files when needed is adequate to efficiently support sound asset management planning.

For utilities that decide they require more sophisticated analyses (e.g. systems that emulate the high-end or strategic options studied in this report at a fairly detailed level of asset hierarchy), Figure 5.2 illustrates that movement toward a fully integrated asset management system can be implemented incrementally over time, based on agency resources and priorities.

APPROACHES TO SUCCESSFUL DATA DEVELOPMENT THAT EMERGED FROM THIS STUDY

Project Data Context

The data gathering for the 1,000-record data samples conducted by the participating utilities for this study represents a microcosm of the process that utilities would encounter in developing an asset register for their full set of assets. Most of the utilities selected assets that would test their ability to gather and use information across their systems (e.g., older and newer segments of the distribution system, a portion of a relatively new WTP, and a portion of an older WTP). The data gathering was divided into two phases. During the initial phase, the utilities spent about 3 months populating the descriptive, financial, and performance information for the sample data sets. These "first-round" data were used to develop an initial set of sample outputs from the project template model. A project workshop meeting was then held to evaluate the state of data development, review preliminary outputs, and to develop strategies for identifying and addressing the most critical of data gaps and other data issues. At the time of this workshop, many of the participants found that there were substantial gaps in their data sets, or that there were data of questionable quality. Table 5.1 illustrates gaps for a portion of the descriptive information for several plant assets for one of the utilities, whereas the unit cost, installation dates, and useful life estimates for these assets are fairly complete.

Table 5.1
Illustration of gaps in data set

Level 4	Plant Item Capacity	Unit of Measurement	Other Descriptor (e.g., Manufacturer)	Unit Cost	Useful Life (Service Life)	Installation Date	Installed Cost (if Available)
AIR HANDLING UNIT	plant				60	1996	$ 22,880
OZONE DESTRUCTION SYSTEM	plant			$ 500	60	1996	
BLUE DIST. TRANSFORMER	15	KVA	CUTLER-HAMMER		60	1996	$ 4,500
OZONE COOLING WATER SYSTEM	plant			$ 500	60	1996	
CLOSED LOOP COOLING WATER SYSTEM	plant			$ 500	60	1996	
180 TON CHILLER PRESSURE GAUGE	plant			$ 500	60	1996	
OZONE CONT OFF GAS ANALYZER 1	plant				60	1996	$ 5,000
OZONE GENERATOR #1	550	LBS/DAY	OZONIA		60	1996	$ 94,212
AMBIENT VAPORIZER1 INLET VALVE	2"		MAGNATROL		60	1996	$ 1,500
AMBIENT VAPORIZER2 INLET VALVE	2"		MAGNATROL		60	1996	$ 1,500
CILLER-COOLSYS FLOW MANOMETER	plant			$ 500	60	1996	
BLUE POWER	plant			$ 500	60	1996	
OZONE CONT OFF GAS ANALYZER 3	0-9.999	PPM	INUSA		60	1996	$ 5,000
OZONE CONT OFF GAS ANALYZER 4	0-9.999	PPM	INUSA		60	1996	$ 5,000
AIR HANDLING UNIT AHU 02	plant				60	1996	$ 22,880
WEST VENT GAS ANALYZER	0-9.999	PPM	INUSA		60	1996	$ 5,000
PLC R10 UPS	plant			$ 500	60	1996	

At the workshop meeting, priorities were developed for refining data and filling data gaps. For example, in the data set above, if the utility thought that the installation dates and estimated unit cost data were not reliable, this was deemed a higher priority than filling data gaps for the descriptors, such as manufacturers. This is because the unit costs and installation dates play an important role in some of the value and performance calculations for the asset management options whereas the descriptors are largely background information. The background descriptive information is useful in understanding the nature of the assets, but it does not play a role in the actual calculation of system value or performance scores. Specific strategies for addressing each utility's priority remaining data needs were identified at the midproject workshop.

Overall Data Recommendations

Following the midproject workshop, each of the participating utilities addressed the priority data issues identified and then submitted a final input data file that was used to develop a final set of output samples for the asset management options. These final outputs for each participating utility are included in the appendix. On the basis of the experience in developing the data throughout the project, the project team identified a number of recommendations related to implementing data strategies for asset management programs. The recommendations were developed to address the fact that most utilities do not have the resources to develop complete asset management databases immediately for all of their assets. The individual strategies listed below might in some cases be implemented individually, but in many cases clusters of these strategies may be implemented by a utility, depending on the data and resource context:

- Address most critical systems first
- Implement first the processes to capture new/replaced assets
- Recognize that there are often substantial gaps
- Explore options for electronic data conversions
- Conduct beta tests of the data-gathering effort to validate both information and the data approach
- Be aware of confidentiality considerations when mapping a data-gathering and integration strategy
- Start at a high level of hierarchy if more-detailed information is not readily accessible with available resources
- Justify strategically the addition of each performance measure
- Look for opportunities to integrate with other systems and processes
- Look for opportunities to centralize information, even partially

Each of these data-related strategy elements, along with other recommendations related to data hierarchies and performance measures, is described in more detail in Chapter 10. The remainder of this chapter addresses in more detail methods that can be used to identify the most critical assets to populate with accurate information in asset management databases.

Determining the Most Critical Assets

Three of the participating utilities (Seattle Public Utilities, the City of Toronto, and the City of Ottawa) included a criticality measure as a supplemental consideration for the data sets that they contributed for this project. By including "criticality" as an additional measure, these utilities were able to assign a criticality score to their assets and then assign a weight to that measure as a contributing consideration of the overall performance score generated by the template model that is used to help identify candidate assets for renewal and replacement activity. By following this methodology, an asset that is ranked high on a criticality scale could be flagged as higher priority for renewal and replacement activity when compared to another asset that scores equally on all other performance measures. While there are more-sophisticated and refined approaches to introducing the consideration of criticality into the analysis, this relatively simple method of introducing a separate additional criticality measure allowed the three interested utilities to evaluate the usefulness of criticality as part of the decision process regarding priorities for renewal and replacement.

As elaborated in their letters of comment in the appendix, the three utilities that tested a criticality measure found that it was useful to take this factor into consideration when determining which assets should be prioritized for renewal and replacement action. Toronto and Ottawa used a fairly simple (low, medium, and high) rating scale developed specifically for this research project to assess the criticality of their contributed assets. Seattle has been working on developing methods to assess the criticality of its water system assets for several years. It used a 10-point scaling system that represented the current scoring of criticality for equipment assets at the time the data were developed for this project, to assign a criticality score for a portion of its assets for which criticality analyses had been completed. The 10-point scoring system, shown below (Table 5.2), provides a more specific differentiation of criticality categories than the more-general three-part categories used by Toronto and Ottawa.

Table 5.2
Seattle Public Utilities equipment criticality

	Code	Description
Equipment is critical to operation of process; criticality of process varies	10	Potential immediate impact to public health and safety: Facility or process is critical to the operation of the system (restoration of process must be immediate), and the individual piece of equipment is critical to the facility or process.
	8	Facility or process is important to the operation of the system, but can be temporarily interrupted for up to 2 days of service; and the individual piece of equipment is critical to the process.
	7	Facility or process is less critical to the overall operation of the system and can be interrupted for more than two days of service, and the individual piece of equipment is critical to the facility or process. (Alternative—the individual piece of equipment is not directly related to the operation of the facility or process but is critical to health and safety, e.g., fire suppression system.)
Individual piece of equipment is less critical to operations, e.g., redundancy	6	Facility or process is critical to the operation of the system and cannot withstand service interruptions, and the individual piece of equipment can be out of service for up to 5 days without impacting the operation of the facility or process.
	4	Facility or process is important to the operation of the system, but can be temporarily interrupted for up to two days of service; and the individual piece of equipment can be out of service for up to 5 days without impacting the operation of the facility or process.
	3	Facility or process is less critical to the overall operation of the system and can be interrupted for more than two days of service; and the individual piece of equipment can be out of service for up to 5 days without impacting the operation of the facility or process.
	2	Equipment is not critical or important to the process and can be down more than 5 days without impacting the operation of the facility or process.
	0	Retired or salvaged equipment.

The higher the number, the more critical the equipment.

The steps that most water utilities have used in the past couple of years to conduct vulnerability assessments for their systems provides a framework that can be considered when developing an overall criticality assessment for the system's assets. Figure 5.3 illustrates the typical steps in conducting a risk assessment for water utility system assets. Determining the risk associated with a facility or set of facility critical assets in a vulnerability context involves a series of evaluations, including:

- Identifying potential threats to the operation of the facility
- Considering the probability that the threat will occur
- Determining the consequences if failure does occur for the asset

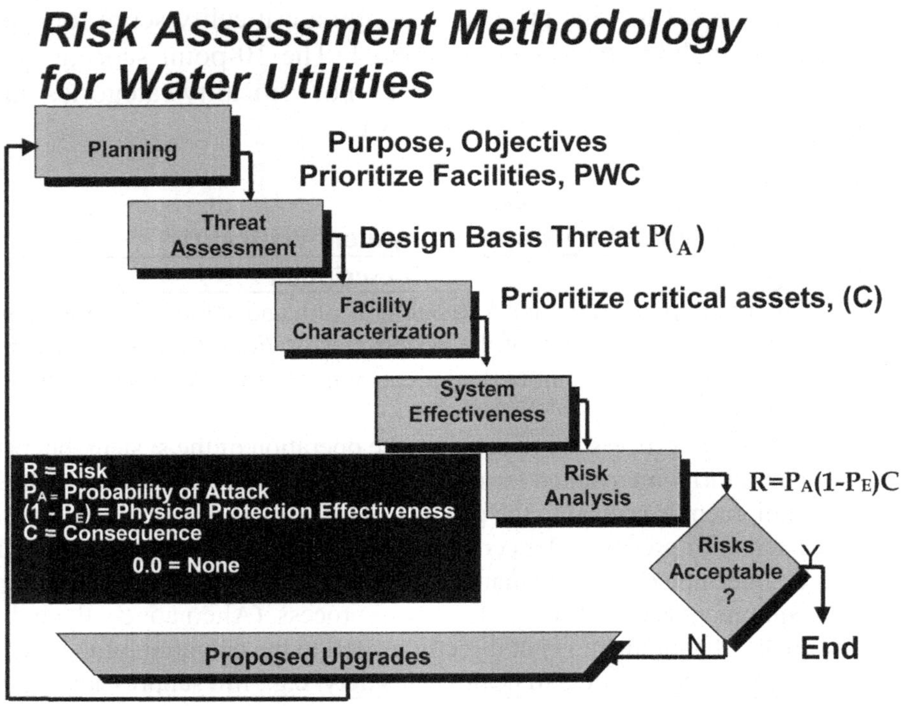

Figure 5.3 Typical steps taken to risk assess a water utility's system assets

The determination of whether the potential risks are acceptable will in part be a function of the results of these technical analyses, but will also take into consideration the risk aversion of the decision-maker and available resources that can be devoted to addressing the most pressing threats. These policy issues and resource constraints are system-specific; therefore, the level of funding implied can vary substantially for two systems that have similar facility characteristics. The four-step process in Figure 5.4 is another way of representing the basic steps in this type of risk assessment:

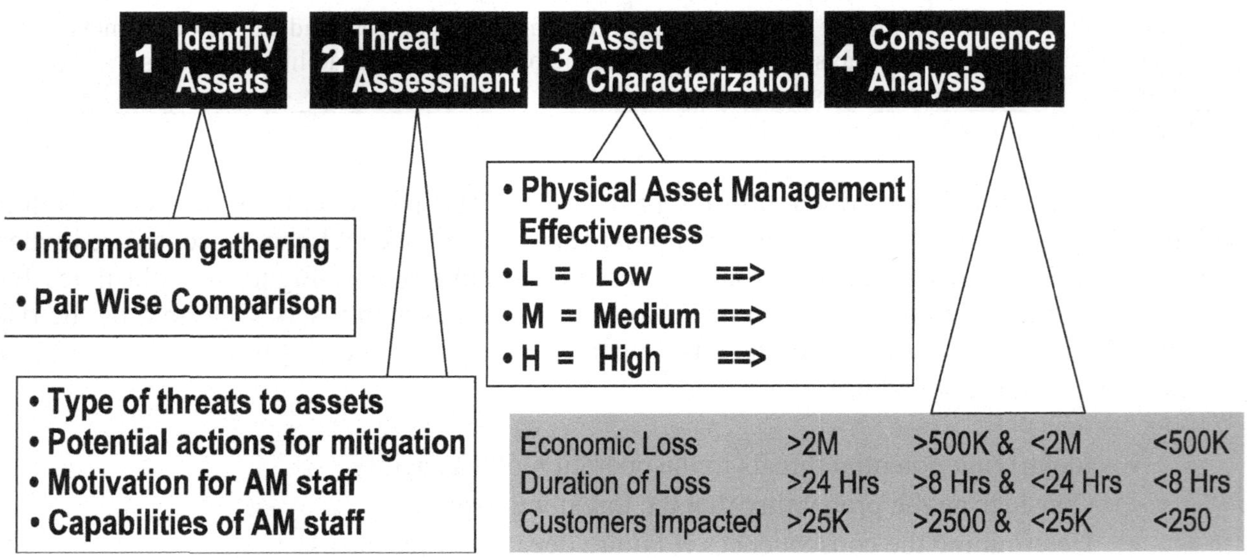

Figure 5.4 Risk for critical assets

Some utilities have developed a matrix approach in assigning a risk-related score to their assets, in which the severity of consequences and the probability of occurance are used as the two axis in a matrix that assigns a risk-related score to the system's assets. Some method of quantifying the risk characteristics along these lines is useful if risk is to be a contributing consideration in asset management planning, as an input to options such as the template model applications for the high-end and strategic options evaluated in this project.

Working through this sort of analyses positions a utilty to answer key strategic questions related to its assets, including:

- Where is significant investment to repair or replace assets required?
- How can different assets impact the utility's goals?
- What assets can create a health or safety issue?
- Which assets can cause an environmental impact?
- Which asset failures can result in a permit violation or other regulatory/legal concern?
- Which assets can create negative publicity or undermine public confidence?

Figure 5.5 illustrates how such a composite risk score might be created, in this case using a maximum score of 100 for assets with the greatest potential consequences and most likely probability of failure:

Likelihood		Frequent		Likely		Occasional			Seldom		Unlikely
	Probability Level	10	9	8	7	6	5	4	3	2	1
Severity	Severity Level										
Catastrophic	10	100	90	60	70	60	50	40	30	20	10
	9	81	81	72	63	54	45	36	27	18	9
	8	60	72	64	56	48	40	32	24	16	8
Critical	7	70	63	56	49	42	35	28	21	14	7
	6	60	54	48	42	36	30	24	18	12	6
	5	50	40	40	35	30	25	20	15	10	5
Moderate	4	40	32	32	28	30	20	16	12	8	4
	3	30	27	24	21	18	15	12	9	6	3
	2	20	18	16	14	12	10	8	6	4	2
Negligible	1	10	9	8	7	6	5	4	3	2	1

Figure 5.5 Sample risk matrix

CHAPTER 6
APPROPRIATE DATA HIERARCHIES

DEVELOPING AN ASSET REGISTER

An asset register is a listing of a utility's assets that contains the data required to manage the assets and is organized into logical groupings. The groupings should "nest" into a hierarchy of larger and larger groups, allowing key information to be summarized at several levels of the hierarchy. Eventually, at the top of the hierarchy is a group that represents the entire facility or system.

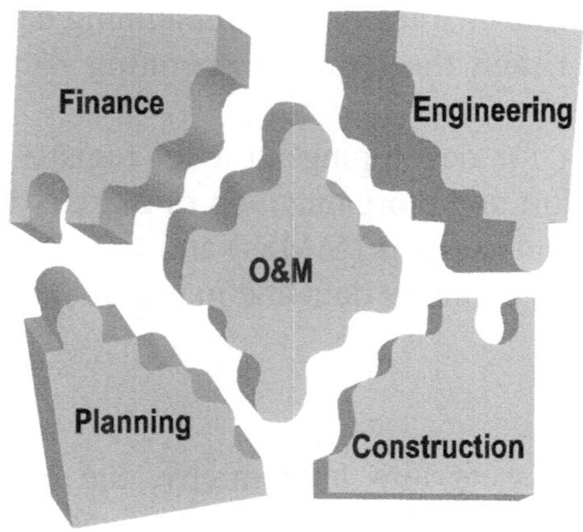

Figure 6.1 Various asset registers that comprise a facility

Typically, utilities have several asset registers. The maintenance department will have a listing of individual assets for tracking scheduled maintenance and repair work. This asset register may be kept manually on card files or spreadsheets or may be kept in a CMMS. In addition, the finance department will have an asset register, frequently called a fixed asset register. The fixed asset register is rarely detailed and lists groups of assets, often organized by the construction contracts under which they were installed. This asset register is used to track capital investments and depreciation for financial reporting. The engineering department may also have an asset register that is linked to design documents and construction records and exists in greater detail than the fixed asset register.

The objective of this project is to evaluate three options for asset management planning and reporting. There are three types of data required for asset management planning and reporting:

- Asset characteristics (e.g., identifier, description, location or system, installation year)
- Planning data (e.g., replacement cost, life expectancy)
- Operating data (e.g., repair costs, break history, water quality history)

To limit the level of effort and to complete this research project within a reasonable schedule, participating utilities selected a limited set of assets to use in evaluating the options. They also chose one of their existing asset registers. In some cases, the assets selected were in an existing register that was already organized into a hierarchy of groupings and contained most of the required data. None of the utilities had an existing asset register that contained all of the data, however.

ORGANIZING ASSETS INTO A HIERARCHY

For long-term planning and reporting, utilities need to be able to easily access required data, and typically those data are not contained within a single database. Utilities may choose to ultimately develop a comprehensive database that fully integrates all of the data needed by all departments. Initially, however, it usually is more practical to implement a "core" database that captures all of the assets organized in a logical hierarchy. This core database would contain basic characteristic and planning data. Periodically, updates would be made to these basic data, and the required operating data would be imported from other databases.

The starting point for developing an asset register is to take advantage of existing databases to minimize effort and reorganization of current data. The greatest level of detail is in maintenance management systems (MMSs) and GIS. These often contain at least some of the asset characteristics and planning data and may also include some operating data.

Asset Management Units

When organizing assets into a hierarchy, the concept of asset management unit (AMU) is helpful. A basic AMU can be an individual asset or component, such as a pipe segment, hydrant, pump, motor, or valve. For some assets, especially if a breakdown by component of characteristics and planning data is not available, it may not be practical or necessary to list the assets by component. For these assets, the basic AMU can be a group of similar assets with the same installation year, replacement cost and life expectancy. For example, all the valves in a geographic area of a distribution system, or all the tanks of a certain type in a treatment plant, may be treated as a basic AMU. In other cases, a functional system whose components can be evaluated together can be treated as a basic AMU for planning and reporting. Even if some data at the component level are available in an MMS or a GIS, this approach is useful when it is cumbersome to evaluate individual components and the system is managed as a single unit. Examples are basic instrumentation systems, structural and architectural elements and landscaping. By identifying these systems as single AMUs, detailed data collection and analysis are avoided. Basic AMUs are then grouped into higher-level AMUs representing geographic units or subsystems and systems.

Defining AMUs

The level of detail used to define the higher-level AMUs should be determined by several considerations. A starting point is to consider any existing groupings that may have been established during the development of a CMMS or GIS, since these already are in use by managers, supervisors, and field staff. However, these groupings often are simply based on geographic areas or plant locations and frequently have only two or three levels. These groupings can be very detailed and often are not be the best for planning, decision-making, communicating, or performance reporting. Creating higher-level AMUs will help provide the appropriate level of detail for all those involved in planning and interested in the performance of the utility.

Stakeholders, such as community groups and elected officials, often want reports organized by neighborhoods or political districts. Executive and higher-level managers often need assets to be grouped into a limited number of systems in order to get a broad overview of performance and budgeting implications. Midlevel managers need information that is sufficiently detailed to allow for thorough analyses of trends and patterns of performance without being overwhelming. Figures 6.2 and 6.3 show example AMU groupings from two utility participants for a distribution system and for a system within a treatment plant.

Distribution System

- *Almaden Valley Pipeline*
  - AVP Vasona - Guadalupe
  - AVP Guadalupe - Shannon
  - AVP Shannon - Calero
- *Snell Pipeline*
  - SPL Colleen - Branham
  - SPL Branham - East Pipeline
- *Central Pipeline*
  - CPL Piedmont Yard to Guadalupe Turnout
  - CPL Guadalupe Turnout to Vasona Pump Station

Figure 6.2: Example hierarchy of AMUs for a distribution system

NaOH Injection System

- *Storage System*
  - Transfer Pump System
  - Tanks to Injection Pumps
- *Injection Pump System*
  - Pump System 1
  - Pump System 2
  - Pump System 3
  - Pump System 4
  - Pump System 5
  - Pump System 6

Figure 6.3 Example hierarchy of AMUs for a plant system

Risk Evaluation

A critical factor in selecting higher-level AMUs is potential risk to the utility. Certain types or asset groups are more critical than others in achieving the goals of the utility. The importance of asset groups should be evaluated against the utility's primary objectives of providing service; protecting health, safety, and the environment; controlling costs; meeting regulatory requirements; and responding to social needs and public perception. The criticality of an asset or asset group can be gauged by determining how severe an impact would occur on achieving the utility's goals if the asset group fails to perform as intended. Poor performance by critical assets presents a higher risk to the utility than the same level of performance by less-critical assets. Examples of high- and low-criticality assets include high-service pumps as compared to local booster pumps, transmission mains vs. laterals, disinfection systems vs. HVAC systems, highly loaded systems vs. systems with excess redundancy. Figure 6.4 shows an example distribution system color coded by the criticality of the water main sections.

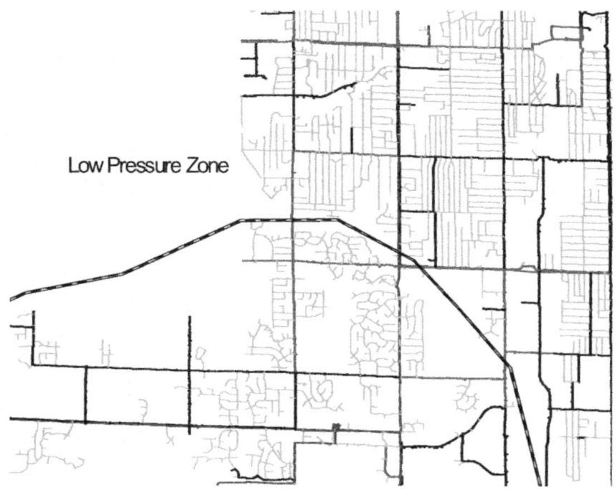

Figure 6.4 Example distribution system highlighting critical pipe sections

HIERARCHIES USED BY PARTICIPATING UTILITIES FOR THIS PROJECT

The general principles for developing a useful asset register hierarchy structure are discussed above. For this project, the participating utilities were constrained by a number of factors, including the current organization of their data and the schedule for completing this project. Their experience is instructive of the kinds of issues facing utilities when they implement a more in-depth process for asset management planning and asset performance tracking.

There is no "right" or "wrong" way to develop an asset register hierarchy. In fact, as a utility's asset management program advances, managers may revise the hierarchy to improve its usefulness. Asset groupings may be disaggregated to provide greater detail where needed or new levels and groupings may be defined to reduce excess detail in analysis or reporting.

Examples of several utilities' approaches are discussed below to show the range of possibilities.

Distribution System Hierarchies

For distribution systems, several utilities had GIS databases that were sufficiently populated for use in developing the asset register hierarchies for this project, while others worked from engineering and construction records.

Toronto Distribution System

Toronto's hierarchy is representative of asset registers developed from a GIS database. Toronto established AMUs based on geography, as did the others who started with a GIS database. The four levels of Toronto's hierarchy are:

Level 1—Entire distribution system
Level 2—Scarborough Pressure Zone 5 (partial)
Level 3—Individual streets within the pressure zone
Level 4—Individual pipe segments (~ 1–30 segments per street)

A total of 818 assets are contained within the portion of the Scarborough pressure Zone 5 selected for this project. Table 6.1 lists several segments and shows how the AMUs and pipe segments are identified within Toronto's asset register. Note that Toronto does not identify hydrants and valves separately from the pipe segments.

Level 3 in the Toronto hierarchy represents City street names. Level 4 represents individual pipe segments wherein each pipe segment has been assigned a unique ID, associated with the centralized infrastructure data registry TIADS (Toronto Infrastructure Asset Data Standard). TIADS contains all pertinent information related to the assets including basic information such as geo-referenced coordinates, date of construction, material type, sizing and condition assessment information. Once fully developed, the City intends for the maintenance management system Hansen, the GIS system (ArcView), and the hydraulic modeling (EPA NET) and future asset management system to be linked through TIADS. Another important distinction is the fact that the pipe segments are based on a "node to node" descritization of the network. In this system, a node represents a location in the network where there is either:

- a tee connection;
- a dead end of a water main;
- a bend or curve; or
- a change in date of construction, material type, or pipe diameter.

Table 6.1
Portion of Toronto Scarborough pressure zone asset register

Level 1	Level 2	Level 3	Level 4
Distribution System	Scarborough Pressure Zone 5 (partial)	ABBOTSFIELD GATE LANE	JAF-L065
Distribution System	Scarborough Pressure Zone 5 (partial)	ABBOTSFIELD GATE LANE	JAF-L066
Distribution System	Scarborough Pressure Zone 5 (partial)	ABBOTSFIELD GT	JAF-L060
Distribution System	Scarborough Pressure Zone 5 (partial)	ABBOTSFIELD GT	JAF-L061
Distribution System	Scarborough Pressure Zone 5 (partial)	ADENCLIFF RD	CDZ-L026
Distribution System	Scarborough Pressure Zone 5 (partial)	ADENCLIFF RD	HBE-L001
Distribution System	Scarborough Pressure Zone 5 (partial)	ADENCLIFF RD	HBE-L002
Distribution System	Scarborough Pressure Zone 5 (partial)	ADIRONDACK GT	DCV-L010

(continued)

Table 6.1 (Continued)

Level 1	Level 2	Level 3	Level 4
Distribution System	Scarborough Pressure Zone 5 (partial)	ALLANGROVE CRES	HBE-L015
Distribution System	Scarborough Pressure Zone 5 (partial)	ALLANGROVE CRES	HBE-L016
Distribution System	Scarborough Pressure Zone 5 (partial)	ALTAIR AVE	HBK-L027
Distribution System	Scarborough Pressure Zone 5 (partial)	ALTAIR AVE	HBK-L032
Distribution System	Scarborough Pressure Zone 5 (partial)	ALTAIR AVE	HBK-L047

Ottawa Distribution System

Another example asset register for a distribution system is Ottawa's. Ottawa also has a GIS database available. Its GIS organized the assets geographically but lists components in more detail, separately identifying hydrants, valves, pipe segments, meters, and service connections. Table 6.2 shows a sample of Ottawa's distribution system asset register. In the Ottawa hierarchy, Level 4 represents a unique ID for individual structures, such as pipe segments, hydrants, valves, meters and service laterals associated with the City's IIMS (Integrated Infrastructure Management System) and/or Geographical Information System (ArcView).

Table 6.2
Portion of Ottawa's Fallingbrook District asset register

Level 1	Level 2	Level 3	Level 4
Fallingbrook	Princess Louise Dr	Hydrant	382037h112
Fallingbrook	Princess Louise Dr	Hydrant	382037h113
Fallingbrook	Princess Louise Dr	Hydrant	382037h118
Fallingbrook	Falwyn Cres	Valve	382038v111
Fallingbrook	Falwyn Cres	Valve	382038v113
Fallingbrook	Falwyn Cres	Valve	382038v115
Fallingbrook	Falwyn Cres	Pipe	P_8238126
Fallingbrook	Falwyn Cres	Pipe	P_8238130
Fallingbrook	Falwyn Cres	Pipe	P_8238133
Fallingbrook	Falwyn Cres	Water Meter	02132012
Fallingbrook	Falwyn Cres	Water Meter	02132014
Fallingbrook	Falwyn Cres	Water Service	S2132012000
Fallingbrook	Falwyn Cres	Water Service	S2132014000

Saint Paul Distribution System

When a GIS database is not available, distribution asset information is most typically found in plat sheet drawings based on construction records. For this project, Saint Paul established an asset register for its selected portion of its distribution system organized around these drawings and the data they contain. A sample is shown in Table 6.3.

Table 6.3
Portion of Saint Paul's distribution system asset register

Level 1	Level 2	Level 3	Level 4
SPRWS	SPRWS-Plat 82	Mains—Plat 82	Raymond to Langford Park
SPRWS	SPRWS-Plat 82	Mains—Plat 82	Raymond to Scudder
SPRWS	SPRWS-Plat 82	Mains—Plat 82	Scudder to Langford Pk
SPRWS	SPRWS-Plat 82	Mains—Plat 82	Intersection of Langford Pk
SPRWS	SPRWS-Plat 82	Mains—Plat 82	Raleigh to Gibbs
SPRWS	SPRWS-Plat 82	Hydrants—Plat 82	Raymond @ Energy Park Dr Sw Cor.
SPRWS	SPRWS-Plat 82	Hydrants—Plat 82	Energy Park Dr @ Everett Nw Cor.
SPRWS	SPRWS-Plat 82	Hydrants—Plat 82	Raymond @ Energy Park Dr Nw Cor.
SPRWS	SPRWS-Plat 82	Hydrants—Plat 82	Raymond @ Blake Nw Cor.
SPRWS	SPRWS-Plat 82	Hydrants—Plat 82	Raymond @ Atty St Sw Cor.
SPRWS	SPRWS-Plat 82	Valves—Plat 82	325' Ww Raymond 8' Out From Where Hyd. #142o Was Located.
SPRWS	SPRWS-Plat 82	Valves—Plat 82	21' We Blake On E Line Of Scudder 10' S & 1' W Of Hyd. #1435 On Blake @ Scudder Ne Cor.
SPRWS	SPRWS-Plat 82	Valves—Plat 82	12' Out From Hyd. #1434 On Blake @ Langford Pk Ne Cor.

Facility Hierarchies

Within asset registers for facilities such as treatment plants and pumping stations, where a populated CMMS is in place, utility participants started with those asset listings.

Ann Arbor Treatment Plant

An example of an asset register imported from a CMMS is Ann Arbor's plant register chosen for this project, shown in Table 6.4.

One can notice that levels 3 and 4 are equivalent in that the data shown in level 3 are the CMMS identifiers for the asset shown in level 4. For the ozone system (level 1), a number of subsystems are identified (at level 2), and the assets are grouped within these subsystems. For the filter press, the level 1 filter press "system" is not divided into subsystems at level 2.

Table 6.4
Portion of Ann Arbor's plant asset register

Level 1	Level 2	Level 3	Level 4
Ozone System	Ozone Destruct	11-380-Ds1-Cc	Catalytic Chamber #1
Ozone System	Ozone Feed/Contact	11-380-Con2-Fga-Control	Flow Gauge Control #2

(continued)

Table 6.4 (Continued)			
Level 1	**Level 2**	**Level 3**	**Level 4**
Ozone System	Ozone Feed/Contact	11-380-Oz-Sam-Pu02	Ozone Sample Pump #2
Ozone System	Ozone Electrical	11-060-Tr-Blue-E	Blue Transformer East
Ozone System	Ozone Electrical	11-060-Tr-Blue-W	Blue Transformer West
Ozone System	Ozone Generation	11-380-Oz-Gn2	Ozone Generator #2
Ozone System	Ozone Generation	11-380-Oz-Gn2-Ps	Ozone Generator #2 Power Supply
Filter Press	Filter Press	11-500-Tk-Hold	Sludge Holding Tank
Filter Press	Filter Press	11-500-Ft-Pu02	Filtrate Pump #2
Filter Press	Filter Press	11-500-Fp1-Pr-Tm	Press #1 Pressure Transmitter
Filter Press	Filter Press	11-500-Fp1-Vl-Sl2918	Press #1 Influent Valve
Filter Press	Filter Press	11-500-Fp1-Vl-Ft4914	Filter Press #1 Prefill Valve
Filter Press	Filter Press	11-500-Fp1-Vl-Ft521	Filtrate Discharge Press #1

Portsmouth Pumping Station

The City of Portsmouth, Virginia, used its engineering and construction records to create the asset register. A portion of the asset register for Portsmouth's Pichkettle Road Raw Water Pumping Station and Intake is shown in Table 6.5.

Table 6.5
Portion of Portsmouth facility asset register

Level 1	Level 2	Level 3	Level 4
Raw Water Supply	Pitchkettle Rd. Raw Water PS & Intake	Miscellaneous	Bond, Insurance, Mobilization, Demobilization
Raw Water Supply	Pitchkettle Rd. Raw Water PS & Intake	Site and Grading	Site Preparation
Raw Water Supply	Pitchkettle Rd. Raw Water PS & Intake	Site and Grading	Drainage Crib Wall
Raw Water Supply	Pitchkettle Rd. Raw Water PS & Intake	Site and Grading	Excavation & Earthwork
Raw Water Supply	Pitchkettle Rd. Raw Water PS & Intake	Structural	Structure & Paving
Raw Water Supply	Pitchkettle Rd. Raw Water PS & Intake	Process Mechanical	Piping & Valves
Raw Water Supply	Pitchkettle Rd. Raw Water PS & Intake	Process Mechanical	Vertical Turbine Pumps
Raw Water Supply	Pitchkettle Rd. Raw Water PS & Intake	Process Mechanical	Sump Pump
Raw Water Supply	Pitchkettle Rd. Raw Water PS & Intake	Process Mechanical	Chemical (KMN04) Feed System

(continued)

Table 6.5 (Continued)			
Level 1	**Level 2**	**Level 3**	**Level 4**
Raw Water Supply	Pitchkettle Rd. Raw Water PS & Intake	Process Mechanical	Floating oil and trash booms
Raw Water Supply	Pitchkettle Rd. Raw Water PS & Intake	Process Mechanical	Static Stainless Steel Intake Screens (6 each)
Raw Water Supply	Pitchkettle Rd. Raw Water PS & Intake	Process Mechanical	Slide Gates
Raw Water Supply	Pitchkettle Rd. Raw Water PS & Intake	Instrumentation	I&C

In this case, a detailed listing of individual assets and components is not available, and the "assets" at level 4 are really groups of assets, or AMUs. Nevertheless, planning projections and performance information can be developed and gathered for these AMUs. Over time, as detailed asset listings become available, these can be incorporated into the asset register at level 5 and if necessary even lower. The level 3 groupings used the construction or bid package information that was available in their records.

Portland Plant

Portland chose to include its NaOH injection system for this project. The injection system (level 1) is divided into three subsystems at level 2, and each of these is divided into two sub-subsystems. These contain between 2 and 40 individual assets. The hierarchy is represented in Table 6.6.

Table 6.6
Portland's asset register hierarchy

Level 1—NaOH Injection System
- Level 2—Storage System
  - Level 3—Transfer Pump System
    - Level 4—Individual assets (4)
  - Level 3—Tanks to Injection Pumps
    - Level 4—Individual assets (10)
- Level 2—Injection Pump System
  - Level 3—Pump Systems (6 systems at this level)
    - Level 4—Individual assets (8 each)
  - Level 3—Conduits (3 systems at this level)
    - Level 4—Individual assets (2 each)
- Level 2—Process Monitoring
  - Level 3—Pump System
    - Level 4—Individual assets (40)
  - Level 3—Monitoring Equipment
    - Level 4—Individual assets (6)

Samples from Portland's asset register are shown in Table 6.7.

Table 6.7
Samples of Portland's plant asset register

Level 1	Level 2	Level 3	Level 4
NaOH Injection System	Storage System	Transfer Pump System	Transfer Pump
NaOH Injection System	Storage System	Transfer Pump System	Check valve on pump discharge
NaOH Injection System	Storage System	Tanks to Injection Pumps	Tank 1
NaOH Injection System	Storage System	Tanks to Injection Pumps	Tank 2
NaOH Injection System	Injection Pump System	Pump System 1	1—Check Valve on alternate supply line
NaOH Injection System	Injection Pump System	Pump System 1	1—Check Valve on pump suction line
NaOH Injection System	Injection Pump System	Conduit 2	C2—Pressure Relief Valve
NaOH Injection System	Injection Pump System	Conduit 2	C2—Flow Meter
NaOH Injection System	Process Monitoring	Pump System	C2—Monitoring Pressure Gage
NaOH Injection System	Process Monitoring	Pump System	C2—Monitoring Pump A
NaOH Injection System	Process Monitoring	Monitoring Equipment	C2—pH Probe/Monitor A
NaOH Injection System	Process Monitoring	Monitoring Equipment	C2—pH Probe/Monitor B

By carefully grouping assets into subsubsystems and systems, Portland has designed an asset register with hierarchies that provide the greatest flexibility in planning and reporting. Very-detailed, asset-by-asset information can be obtained by analyzing the data at level 4. Supervisors and line managers can review performance results at levels 2 and 3 and focus decision-making about maintenance programs and renewal and replacement. Executive managers, budget officials, regulatory agencies and outside observers can gain sufficient understanding of plant performance and budget proposals by reviewing information at level 1.

CONCLUSIONS

A number of conclusions can be drawn from the experience of the utilities in developing asset registers and organizing them into useful hierarchies for their asset management programs. An ideal asset register is structured so that there are levels of detail and groupings of assets that meet the needs of everyone who uses the data and results. These include operators, engineers and planners, up through the organization's middle and executive managers, to the Board or Commission, elected officials, public organizations, and regulatory agencies. While these interests may change over time, there usually are logical groupings by geographic boundaries and process systems that are useful starting points. A review of risk considerations when constructing an asset register hierarchy will allow critical assets and systems to be grouped and highlighted during planning and reporting. All this implies, therefore, that there is no uniform hierarchy that should be adopted by all water utilities.

Most importantly, however, when a utility implements an asset management program, it should begin with what resources and data are available. If a well-populated GIS or CMMS already exists, the detailed databases these contain should be used. Many of these systems identify groupings that were developed when they were implemented and these should be used when appropriate. If a manual maintenance management system, engineering or construction records, or a fixed asset register prepared for financial reporting are what are available, these are often adequate to begin the process of more systematic and thorough planning and reporting. The emphasis when advancing an asset management program should be on adopting existing databases rather than delaying improved management of the assets while data is gathered and reorganized.

Additional related findings and recommendations are contained in Chapters 9 and 10, respectively.

CHAPTER 7
DEVELOPING AN APPROPRIATE EVALUATION FRAMEWORK

Considerable effort was expended by the participating utilities in developing and then completing evaluations of the asset management options studied in this project. While the sample outputs (both tables and figures) contained in the appendix are useful in providing readers samples of what they could expect to see as outputs if the various levels of asset management planning were implemented, the evaluations and insights of the participating utilities, who have in most cases contributed more than 200 hours of in-kind support to the development and evaluation of the report options, provide additional valuable context for considering the usefulness of the various tables, figures, and options studied.

OVERVIEW OF THE EVALUATION COMPONENT OF THE PROJECT

The evaluation process developed for the project consists primarily of the following elements:

- A **scorecard**, completed by each of the participating utilities, which assesses the value of and difficulty in implementing the asset management options, the performance measures studied, and each of the tables and figures developed to illustrate the outputs for the three asset management options studied. Each utility's completed scorecard is included in the appendix; the scorecard ratings are summarized and analyzed in Chapter 9, Findings. A copy of the blank scorecard form is included at the end of Chapter 2. As shown in the blank form, in most cases there is a separate evaluation provided regarding the value and difficulties in gathering information for water treatment plant and distribution system assets. This separation was suggested by the participating utilities on the basis of their experience throughout the project finding that there were noticeable differences in their experience with regard to plant vs. distribution system assets. This is borne out by the differences in the resulting evaluations, summarized in Chapter 9.
- A **letter of comment** has also been provided by each participating utility to provide additional perspectives on the value of the options and outputs. The letters, which are also in the appendix, contain additional insights from the partners, based on their considerable effort in developing their sample data sets and their active participation in the development and evaluation of the options and outputs studied in this project.
- **Additional analysis and commentary** regarding the usefulness of the options and outputs by the research team is provided in Chapter 9.

AUDIENCES FOR ASSET MANAGEMENT–PLANNING INFORMATION

For the evaluation of the asset management options and outputs to be useful, the audiences for the information developed need to be carefully considered. During the execution of this project, substantial attention was given to the identification of the target audience(s), which were broadly divided into two groups:

- **Utility Management Decision Makers**—Includes the utility study participants themselves, which cover a broad spectrum of agency roles, including two utility directors, several finance department representatives, several IT department representatives, and several engineers who have been given responsibility within their utilities for developing or managing asset management programs. In addition, this audience includes others within the respective utility organizations whose support is needed in order to recommend proactive asset management programs to the agencies' governing bodies and other external stakeholders.
- **Utility Governing Bodies and Other External Stakeholders**—Includes all audiences outside the utility's internal staff. In some cases, there is a fairly small group of audiences outside the agency staff, including the governing board, whose support is necessary to secure funding for identified, recommended asset management programs. In other cases, there is a significant number of key actors outside of the agency's staff whose support is critical to successful implementation of an asset management program.

The remainder of this chapter lists the key stakeholder groups that have been identified by the research team and the participating utilities during workshops, conference calls, and other project communications. During project workshops, specific key individual stakeholders were identified to help the utilities focus their evaluation scorecards and letters of comment on the people whose support would be essential to secure funding for recommended asset management programs. In the sections below, this stakeholder identification process has been generalized to identify the stakeholder groups within and outside of the utility organization whose support may be needed. The information needs for these stakeholder groups, and the kinds of output information and performance metrics deemed of value to these groups, are detailed in Chapter 9.

Internal Utility Agency Stakeholder Groups

During the course of the project, the following internal stakeholder groups were identified as potentially critical to securing support for funding for recommended renewal and replacement programs:

- Senior agency management
- Finance department
- Engineering department
- Construction management groups
- CIP planning director
- IS/IT managers (to understand needs and help with required data management)
- Trade supervisors
- Maintenance managers
- Operations managers
- Risk managers
- Procurement department
- Rate department

- Plant or distribution manager
- General rank-and-file employees (ambassadors within the community of relatives, neighbors, friends)

External Stakeholder Groups

During the course of the project, the following external stakeholder groups were identified as potentially critical to securing support for funding for recommended renewal and replacement programs:

- Mayors, city council members
- Board of directors or other governing bodies
- Retail customers
- Large commercial/industrial customers
- Large wholesale customers (e.g., neighboring communities that buy bulk water)
- The media
- Bond-rating agencies, underwriters, bondholders
- Regulatory/rate commissions

CHAPTER 8
RELATIONSHIP TO RELATED PLANNING ACTIVITIES

Chapter 7, which addresses the evaluation process conducted for this project, identifies a significant number of internal and external stakeholders who are likely recipients of information developed for an asset management program. Many of these recipients will be responsible for a variety of utility planning processes which may be impacted by the results of asset management planning. Some of the primary interfaces are described below.

RELATIONSHIP TO STRATEGIC PLANNING

Maintaining treatment and distribution system facilities in sound working order is a core business for any water utility. As such, the development of plans and budgets to support appropriate levels of investment in renewal and replacement (R&R) of aging assets is a core element of any sound strategic business plan. Even if a utility has strong interest in branching out into other businesses, it cannot neglect its core assets.

Prior Research Foundation studies address the elements of a strategic planning process in more detail. The tailored collaboration project Development of a Strategic Planning Process (AwwaRF, 2002), identified a five-step process for water utility strategic planning:

1. **Lay the foundation.** A solid foundation is essential for a viable strategic plan. Considering the political realities within which utilities operate, this step involves working with policy boards to define baseline conditions, policy directions, and boundaries. As strategic planning is integrated into the annual planning cycle, this step will involve revisiting policy directions based on changes in conditions, policy board members and priorities, and stakeholder expectations.
2. **Identify options.** Working with stakeholders, the utility assesses community priorities and confirms its vision, mission, and goals in light of step 1. Strategic opportunities begin to take shape in this step.
3. **Evaluate options.** Resource investment strategy options are clearly defined, an evaluation model that reflects stakeholder priorities is constructed, and options are evaluated and scored.
4. **Develop strategic plan.** Ranked strategic investment options are grouped into "investment portfolios" and prioritized to form a complete strategic plan, recognizing synergies across operating units.
5. **Plan for implementation.** Implementation and contingency planning is conducted to help assure the strategic plan is interwoven with the utility's business processes. Final approval is solicited from the policy board, and implementation begins. A system for monitoring progress is put in place to facilitate annual updates of the strategic plan.

For almost every water utility, asset management is a key element of this strategic planning process. As shown in Figure 8.1 from the tailored collaboration project, asset management is one of the key investment options that is typically considered:

Figure 8.1 Strategic investment options

The typical utility business cycle identified in the strategic planning project is illustrated in Figure 8.2.

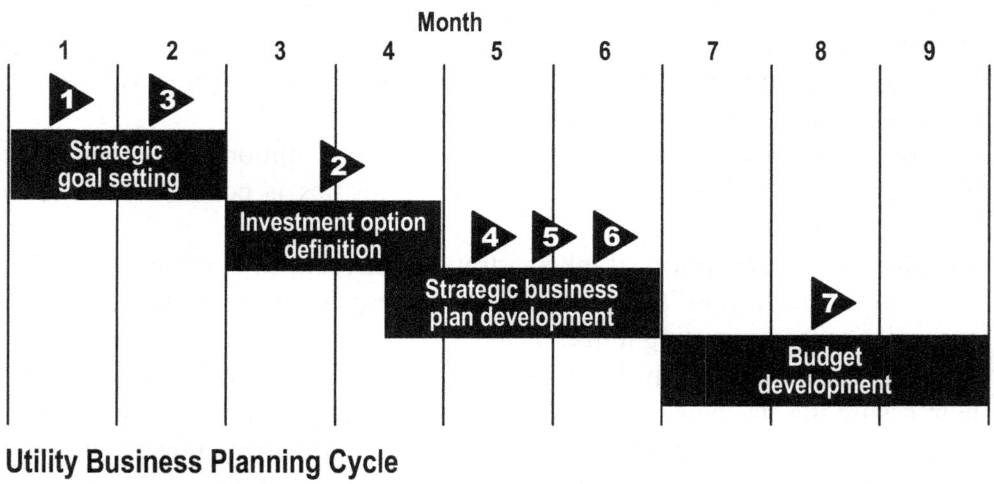

Figure 8.2 Utility business planning cycle

Asset management and its implications for sound financial and business practices for water utilities should be considered and integrated throughout this process, from the initial goal setting through the development of plans and budgets. This is particularly true in light of industry studies, summarized in Chapter 1, that have documented the substantial backlog of renewal and replacement investments faced by most water and wastewater utilities.

For example, when the Honolulu Board of Water Supply established a value model as part of its participation in the tailored collaboration project, asset management considerations such as "optimize assets" and "system reliability" were identified as key strategic goals. Increasingly, utilities are defining their asset management programs in a manner that encourages linkages across other functions. For example, in her appended letter of comment, Liz Kelly, Director of the Strategic Operations Division at Seattle Public Utilities, says: "At SPU, we are in the midst of implementing a comprehensive strategic asset management program. We have adopted a very broad definition of asset management that combines risk management, thoughtful capital project and program implementation, and maintenance of practices and strategies." Her letter goes on to detail more than a dozen elements of this broadly defined program.

RELATIONSHIP TO FINANCIAL PLANNING

The development of capital and operating-budget assumptions for renewing, rehabilitating, and replacing aging system assets is a key challenge for most water utilities. Historically, the level of funding has often been driven by regulatory requirements, limited budget allocations, and what customers and governing bodies will approve or are perceived to be willing to approve. The development of proactive asset management programs that provide technical analyses of the age, condition, useful lives, and resulting needs for reinvestment add important objectivity into what might otherwise be a subjective discussion of R&R expenditure needs.

As such, the tables and figures that result from asset management programs, such as the examples for the participating utilities illustrated in Chapter 3 and in the appendix, serve as key inputs to a utility's financial planning processes that, depending on system size and the stage of development, can include:

- **Operating Budget Projections**—Since some R&R needs are relatively small expenditures that may not be capitalized, asset management planning can identify key inputs to a utility's operating budget projections. Depending on the data platforms used and on how data are maintained, data may flow to and from a utility's CMMS or other tools used to track operation-and-maintenance requirements.
- **Capital Budget Projections**—Implementation of an asset management program along the lines of the basic option evaluated for this study can help to set overall R&R funding levels in the capital budgeting process. Implementation of the high-end option or the strategic option will help to identify specific projects for inclusion in the capital budget for future years.
- **CIP Programs**—R&R needs and the projects created to address them serve as key inputs to a utility's CIP program development.
- **Bonding/SRF Program Requirements**—Asset management studies can provide key supporting documentation for a municipal bond market prospectus or as part of an application for State Revolving Fund (SRF) loans. Quantification of the system's ongoing R&R needs, backed up by objective, replicable analyses, helps to create a credible case for funding support. This can help address the uneasiness that lenders may have due to the historical gap in knowledge of systems and their true R&R

needs. The appendix contains several contributions from Standard & Poor's (S&P) that address the increasing importance of realistically estimating future system needs in utility system credit analyses. In his appended white paper, *Credit Implications of Rate Structure and Rate Setting for U.S. Municipal Water-Sewer Utilities*, James Wiemken, S&P's sector leader for municipal water and sewer utility ratings and state revolving fund ratings throughout the United States, points out that

> True rate stability and transparency assumes that a system's current and likely future needs have been measured and are relatively known. The average increase in rates to be targeted over the next decade cannot be known without some idea of the cost pressures a utility may face, and without an honest effort to estimate these needs, it will be extremely difficult to educate and inform ratepayers.

RELATIONSHIP TO FINANCIAL REPORTING AND FIXED ASSET SYSTEMS

Fixed asset and financial reporting systems for utility systems have traditionally been developed as freestanding applications, often by the Finance or Accounting departments within the utilities or the related municipal government. Until recently, little effort has been made to integrate these systems with the databases that are used to plan for and manage the utility systems, such as GIS and CMMS. As detailed in Chapter 1, the release in June 1999 of Statement 34 by GASB is one strong example of the trend to recognize opportunities to link financial planning, management, and reporting functions.

The asset management option identified in Statement 34 is not directly applicable to many water utilities because many are organized as enterprise funds that may continue depreciating their assets to satisfy financial-reporting requirements. However, to the extent that there are valuable insights that result from GASB's research in developing Statement 34 that may benefit water utilities in their asset management planning, those insights should be captured. The performance measure–based options for this project (the high-end and strategic options), address the spirit of GASB's Statement 34 since they produce definable, replicable approaches for identifying the condition of various utility system assets. The participating utilities' assessment of the value of the high-end and the strategic options for this study illustrates the interest of diverse water utility systems in moving toward asset management systems that address many of the issues considered by GASB when it formulated the asset management option.

Utilities that choose to implement any of the options studied in this report (basic, high-end, or strategic) would create databases to store information that would have much in common with information from databases required to support the fixed asset systems and financial reporting systems. Common data elements would include asset identification codes, installation dates, and cost information.

Even in the situations where the financial reporting is based on the depreciation of original asset costs, and the asset management systems are based on replacement values (basic option) and performance measures (the high-end and strategic options), it would be valuable to integrate the management of common data elements (e.g. installation dates and asset identification codes). Such an integration would serve several useful purposes, including:

- Providing a single point of electronic entry for this common information for new assets as new systems are added or as assets are retired and replaced
- Enabling an additional validation/verification point for both the asset management and the fixed asset systems, by conducting an exceptions-reporting process to identify assets included in one but missing from the other system

The interfaces with financial reporting and fixed asset systems is described in further detail in the letter report by American Appraisal Associates, which is included in the "Other Research Team Contributions" section of the appendix. In his appended letter report, Paul Gruenwald, Vice President of AAA and Managing Principal for the firm's Milwaukee Property Appraisal Services Group points out:

> From experience in appraising Water Utilities, we find that what appears in the capital asset ledger maintained by the Finance Department is different than what is actually in the plant. This is because there are normally no established procedures for plant personnel who are familiar with what is occurring in the plant to report retirements and additions to the Finance Department. If an electronic interface could be established that would allow plant personnel to automatically report to the Finance Department assets that are being constructed, replaced, or disposed of, it would improve the accuracy of the capital asset ledger, reduce the manual effort in the Finance Department to capture this information, and allow the Finance Department to report better information.

CHAPTER 9
FINDINGS

Given the diverse nature of the technical and management studies conducted as part of this project, there are findings in a number of areas:

- Overall value of the asset management options studied
- Overall value of performance measures studied
- Data management challenges
- Cultural and organizational challenges increase for higher-end options
- Added value often validates movement toward high-end and strategic options
- Strategic use of data hierarchies
- There are many relevant stakeholder groups, many with specialized communication needs

Research team and participating utilities' overall findings are discussed below. Where appropriate, findings from the scorecards completed by the utility participants and comments from letters of comment developed by the participating utilities. The individual scorecards completed by each utility participant and the letters of comment in full are included in the appendix.

OVERALL VALUE OF ASSET MANAGEMENT OPTIONS STUDIED

A key finding of the study is that all three of the asset management options studied provide, when compared with no action, substantial value in addressing the needs of managers and for stakeholder outreach. On a five-point scale, in which 1 represents not very valuable, 3 represents somewhat valuable, and 5 represents very valuable, the participating utilities rated the basic option 3.3 for plant assets and 3.5 for distribution system assets. The high-end option was rated 3.8 for plant assets and 4.0 for distribution system assets, whereas the strategic option was rated 4.0 for both plant and distribution system assets.

The basic option provides information that can be used to define overall renewal and replacement needs. During project workshops, many of the utilities indicated that this level of information is a very useful starting point in moving from underfunded asset management programs to programs that provide for adequate resources. Particularly in the early years of an asset management program, this level of information may be adequate to guide financial commitments, as there are often easily identifiable projects that should be executed, which can be justified and supported by setting aside the funds identified as the target renewal and replacement investments. Several of the utility participants also commented that, because most of the information to implement this option is fairly accessible, this option could be a good starting point for utilities that currently don't have an organized asset management program. Key limitations mentioned by some of the utility participants for this option are the challenges in determining appropriate service lives and the lack of consideration of asset criticality/significance in the analysis framework.

The high-end and strategic options provide substantial additional insight into the assets that should be targeted for action. When appropriate performance measures are identified and prioritized, the research team and utility participants found substantial additional benefit in implementing these types of asset management options. As the "low-hanging fruit" projects that may be implemented during the initial years of an asset management program are completed, the asset-specific performance and condition information available through the high-end and strategic options was found to provide substantial additional value to aid management decision-making, and making the case to stakeholder groups.

The high-end option was seen by some of the utility participants as particularly useful in setting the appropriate asset renewal strategy for many assets (e.g., groups of assets within a single geography) and as an indicator of condition. The strategic option was rated somewhat higher than the high-end option. Several of the participants mentioned as the basis for this difference the fact that the strategic option enabled dynamic funding evaluation.

The substantial data requirements for both the high-end and strategic options were cited as important considerations that need to be taken into account when one decides on an implementation strategy, both in terms of timing and the level of detail at which these options would be implemented.

OVERALL VALUE OF PERFORMANCE MEASURES STUDIED

Value of Commonly Studied Measures

Overall, the study team found the use of prioritized performance measures to be a valuable way to characterize assets and guide renewal and replacement investments. The ability to set system-specific weights to the performance criteria and the ability to set system-specific scales in assigning value points were viewed as useful features that helped to address likely decision-maker and stakeholder interests more directly. More than half of the participating utilities took advantage of these features by assigning customized weights or scales for their systems. In his appended letter of comment, the San Francisco Public Utilities Commission's Finance Rate Administrator William Laws explained the basis for weighting some measures in relation to access to reliable information and system priorities:

> The Public Utilities Commission gave greater weighting to the criteria that were currently measured or considered easy to measure. The service factors important to the Public Utilities Commission's customers are water quality and service reliability. Consequently, those factors together with scheduling and performing preventive maintenance and age versus service life received the majority of the weighting.

Clifford Jamile, Manager and Chief Engineer of the Honolulu Board of Water Supply, explains in his appended letter of comment that,

> In moving forward with full implementation of an asset management program, I am certain that we would customize and adapt the weightings and scaling factors to reflect our actual experience in working with our assets, and by observing the results of several rounds of analysis using a decision-support tool.

As detailed in Chapter 2 (methods), the study team selected seven out of approximately 50 candidate performance measures for common study based on perceived value and likely access to information to populate the data sets within the several-month period allowed for data development during the project. Midway through the project, several of the performance measures were redefined. This was done in some cases to provide additional clarity in how information was categorized and in others to improve the value of the measures on the basis of the review of preliminary rounds of output information. Based on the final set of seven commonly studied performance measures, the service reliability measures for both distribution system and plant assets were deemed by the participating utilities to be among the most valuable as aids to management decision-making regarding which asset management strategies to recommend to the utility-governing boards and stakeholder groups. Below is the participating utilities' average ranking of the performance measures on a five-point scale, with 5 representing maximum value[1]:

Service reliability-main breaks (distribution system)	4.4
Service reliability-interruptions (plant)	4.3
Maintenance activity (plant)	4.3
Age vs. service life (distribution system)	4.0
Maintenance activity (distribution system)	3.9
Water quality (plant)	3.8
Preventive maintenance (plant)	3.8
Age vs. service life (plant)	3.8
Water quality (distribution system)	3.7
Service reliability-interruptions (distribution system)	3.5
R&R status (distribution system)	2.8
R&R status (plant)	2.7
Preventive maintenance (distribution system)	2.6

As seen in the ranking above, the R&R status measures and the preventive maintenance measure for the distribution system were ranked the least valuable as aids to management decision making. Several of the letters of comment developed for the project explain some of the differences in the rankings of the performance measures for plant and pipe assets. For example, in her appended letter of comment, Christine Meyer, Information Services Manager for Saint Paul Regional Water Services explains different rankings for the "Service Reliability—Interruptions" measure: "For the collected data of this study, SPRWS felt this performance measure is much more applicable to Plant assets and that the scoring emphasis should reflect this. Plant processes are redundant and unplanned interruptions do not have serious impact on service reliability."

In terms of value in making the case to utility governing boards and stakeholder groups, several of the performance measures rank very highly. As shown in the ranking below, seven of the measures ranked above 4.0 on a five-point scale, with service reliability, water quality, and age vs. service life measures all ranking above 4.0. The measures deemed of least value in

[1] The listing shown here is the average (mean) rating by the utilities that rated these measures. For most of the performance measures, there were at least some ratings as low as '1' or '2', and some as high as '4' or '5'; the same is true for the other lists of average ratings reported later in this chapter. The specific ratings of each measure by each of the utilities were recorded on scorecards that are included in the Appendix; interested readers can therefore review the ratings of specific performance measures by individual utilities by reviewing the scorecards.

making the case to stakeholders and utility-governing boards are the R&R status and preventive maintenance measures.

Service reliability—main breaks (distribution system)	4.8
Water quality (distribution system)	4.5
Age vs. service life (distribution system)	4.4
Service reliability—interruptions (plant)	4.3
Service reliability—interruptions (distribution system)	4.3
Water quality (plant)	4.2
Age vs. service life (plant)	4.2
Maintenance activity (plant)	3.5
Maintenance activity (distribution system)	3.5
R&R status (plant)	2.9
Preventive maintenance (plant)	2.9
Preventive maintenance (distribution system)	2.9
R&R status (distribution system)	2.8

A number of the participating utilities cited stakeholder interest as very important when decided which performance measures are most useful. For example, in his appended letter of comment, James Spacek, Director of Public Utilities for Portsmouth, Virginia's Department of Public Utilities says:

> The performance measures believed to have the most value are directly related to stakeholders and therefore have high (and generally immediate) value. For example, we've rated water quality and service reliability, particularly at the water treatment plant, as very valuable. Our perception is that this will be both immediately noticeable and could immediately impact the health and welfare of our customers.

Several of the participants commented directly on the reasons that some measures ranked differently for stakeholder outreach than for other purposes. For example, Clifford Jamile, Manager and Chief Engineer of the Honolulu Board of Water supply, indicated in his appended letter of comment that

> In our view, the maintenance activity and preventive maintenance measures, while very valuable for management decision-making, are of somewhat lesser value in stakeholder communications; these particular measures are probably not of as much interest to a lay audience as some of the other measures, such as service interruptions and water quality violations.

Additional Measures of Value

A few of the participating utilities gathered information for supplemental measures for inclusion in their data sets for the high-end and the strategic options. Several were found by the utilities studying them to add substantial value to the analysis:

- **Hydraulic Performance**—Toronto included a measure of hydraulic performance in its data set that was found to provide additional insights when considering which segments of the distribution system should be targeted for renewal or replacement action. Many of the participants thought that such a measure would be valuable. Such a measure was not included in the commonly studied measures because most of the participating utilities did not have access to such information. In his assessment of the usefulness of this measure, Michael D'Andrea, Manager of Infrastructure Asset Management for Toronto, indicates in his appended letter of comment that "With further refinement in the approach, and coupled with watermain break rates, it could be used to determine whether a pipe segment should be rehabilitated (e.g., cleaned and lined and cathodically protected) or replaced/upgraded."

- **Criticality**—Seattle, Toronto, and Ottawa included a measure of criticality in their evaluations. These measures assign a score that takes into consideration what happens if the use of an asset or asset group is lost and the consequences of asset failure. The participants who included such measures found that they provided useful additional information when targeting assets for action in the high-end and strategic options. Many of the other participants also thought that a criticality measure would be valuable. Such a measure was not included in the commonly studied measures because most of the participating utilities did not have access to such information. In her appended letter of comment, Liz Kelly, Director of Seattle Public Utilities' Strategic Operations Division, offers the following assessment on the importance of this measure:

> Criticality information is of particular importance to SPU....At SPU, we rate the criticality of each asset on a one to ten scale—higher numbers on the scale indicate that failure of the asset would have immediate impact to public heath and safety, or it is critical to the operation of the system.

DATA MANAGEMENT CHALLENGES

At the outset, it was anticipated that there would be significant challenges in collecting information required to support the asset management options being studied. This was due in part because the required information, particularly that needed to support the high-end and strategic options, is, if stored electronically, often stored in separate databases that include the utility's fixed asset database, computerized maintenance management system, budget and financial recording systems, geographic information systems, and other free-standing databases and spreadsheet files. Thus the target data file was limited to 1,000 records in order to provide some boundaries to the data-gathering efforts within the project schedule.

The data management challenges proved even more significant than what had been anticipated at the outset of the project. During the first project workshop, when the options for study were selected and performance measures defined, a third of the participating utilities indicated that they would likely have fairly easy access to most of the identified information in electronic formats. The rest of the participants anticipated challenges in gathering at least some of the financial and performance information identified for study. Over the course of the project, almost all of the participating utilities encountered substantial data management challenges for at least some of the required input data. On average, each of the participating utilities spent approximately 200 hours developing their input data files and contributing to the development of the options and performance measures through their participation in workshop meetings, conference calls, and related activities. Typical of the data management challenges encountered by the participating utilities is this example from the appended letter of comment from William Laws, Rate Administrator at the San Francisco Public Utilities Commission:

> The Public Utilities Commission does not currently link its asset registry to its computerized maintenance management system. Both systems were developed as stand-alone systems and data currency is not maintained between the two systems. Considerable effort was required to match the performance data in the computerized maintenance management system to financial data in the asset registry.

Overall, the participating utilities assigned a score of approximately 3 to the difficulty in integrating data across multiple databases for their assets (with 1 being very difficult and 5 not very difficult). In terms of data management challenges for the three asset management options, the challenges associated with the high-end and strategic options were found by the participating utilities to be significantly greater than those associated with the basic option:

- For plant assets, data management challenges associated with the basic option were rated 3.9 on a five-point scale (with 1 defined as very challenging, 3 defined as somewhat challenging, and 5 defined as not overly challenging). The data management challenges associated with plant assets for the high-end option were rated 2.8, whereas those for the strategic option were rated 2.4.
- For distribution system assets, data management challenges associated with the basic option were rated 4.2 on the same scale. The data management challenges associated with distribution system assets for the high-end option were rated 2.5, whereas those for the strategic option were rated 2.1.

In her appended letter of comment, Christine Meyer, Information Services Manager at Saint Paul Regional Water Services, describes some of the specific challenges encountered when populating the data sets for the project, such as:

> …I learned that maintenance or replacement of hydrants could also involve possible maintenance or replacement of valves and that the cost information was commingled for these two asset types. Through interviews with distribution foremen, I was able to manually separate out the labor and materials associated with this activity.

In terms of difficulty to obtain information about the specific performance measures studied, the R&R status information and preventive maintenance information for plants was deemed the most difficult. Those measures were rated 2.8 on a five-point scale, in which 1 represents very difficult, 3 represents somewhat difficult, and 5 represents not very difficult. As shown in the listing below, the measures deemed least difficult to obtain are service reliability (main breaks) and water quality violations for both plant and distribution system assets.

Service reliability—main breaks (distribution system)	4.4
Water quality (plant)	4.3
Water quality (distribution system)	3.8
Age vs. service life (distribution system)	3.7
Service reliability—interruptions (plant)	3.3
Maintenance activity (plant)	3.1
Service reliability—interruptions (distribution system)	3.0
Preventive maintenance (distribution system)	3.0
Age vs. service life (plant)	3.0
Maintenance activity (distribution system)	2.9
R&R status (distribution system)	2.8
Preventive maintenance (plant)	2.8
R&R status (plant)	2.8

Some specific points made by the participating utilities related to data gathering and management issues discovered through the conduct of this research project include:

- Some participants identified the need to refine, revise, or improve the quality of data in several existing databases and other electronic data sources (CMMS systems, GIS systems, fixed asset records).
- Some participants identified the need to recode water quality data in order to use it properly.
- Several participants identified the need to change the way that they collect and manage financial information. In some cases, original cost information was not available at the detail needed for asset management analysis. In other cases, historical expenditures on renewal and replacement were not broken down by asset groups, or identifying whether expenditures were reactive or preventive in nature was difficult.
- Several participants found that it was difficult to estimate useful service lives for their systems with reliability, because actual service histories are not being recorded in a way that could be used to inform future planning and decision-making. For example, one participant indicated that there was an absence of "run-time" data, making it difficult to make informed decisions on system replacement needs.
- Some participants found surprising gaps in the consistency of information (e.g., information was relatively complete in some geographic areas, but there was substantial, unexpected gaps in other geographic areas).

CULTURAL AND ORGANIZATIONAL CHALLENGES INCREASE FOR HIGHER-END OPTIONS

The participating utilities were also asked to evaluate the degree to which implementing the three asset management options would pose organizational and cultural challenges within their respective organizations. This evaluation was done on a five-point scale in which 1 designates many anticipated challenges, 3 designates some anticipated challenges, and 5 designates few anticipated challenges.

The basic option—which is based on useful life concepts that are extensions of the depreciation analyses conducted by many utilities as part of their fixed asset reporting and financial statements—was viewed as posing the fewest cultural and organizational challenges. Some of the participating utilities, such as Portsmouth, Virginia, currently employ asset management approaches and systems that emulate most aspects of the basic option. For these utilities, there would be very little in the way of organizational or cultural challenges in implementing the basic option. Overall, the participating utilities rated the basic option 3.9 for plant assets and 4.1 for distribution system assets in terms of the degree of cultural and organizational challenges.

The high-end and strategic options were rated as having much more substantial organizational and cultural challenges. This is not surprising, because most utilities do not currently employ asset management systems based on performance measures. In addition to addressing the substantial data-gathering and data integration requirements to implement these options, many of the utility representatives indicated that the transition to a culture in which performance measures are weighted, routinely tracked, and relied upon as a significant element in project decision-making would represent a substantial change from current practices. Overall, the high-end option was rated 2.8 for plant assets and 2.9 for distribution system assets. The strategic option was rated even more challenging than the high-end option (2.4 for plant assets and 2.3 for distribution system assets).

Despite the substantial organizational and cultural challenges identified, particularly for the high-end and strategic options, the participating utilities saw evidence that diverse groups within their utilities were prepared to work together to implement more-sophisticated asset management programs if useful results could be proven. Some participants mentioned that a cadre of people across organizational lines worked closely together to produce the input information required for the project and indicated that they would be willing to make compromises to realize an enterprise solution rather than to continue to work as separated niche operations. Some participants mentioned that support for enhancing asset data and planning in their organizations has "bubbled up" from the grass roots of the organization, with staff having made the case to senior management of the need to secure resources to enhance these efforts.

ADDED VALUE OFTEN VALIDATES MOVEMENT TOWARD HIGH-END AND STRATEGIC OPTIONS

The value of moving toward the high-end and strategic options depends on a variety of factors, including the complexity of the assets and the number of variables considered important in a utility's decision-making environment. In some cases, participating utility representatives indicated that the high-level information regarding renewal and replacement funding needs produced by the basic option would be considered a sufficient basis to support funding by the governing bodies and stakeholder groups. In other cases, the representatives indicated that their governing boards would expect information such as that produced by the high-end or strategic options before R&R budgets would be approved.

Through their efforts to populate the 1,000-record data sets developed for this project, the participating utilities were well aware of the substantial additional data-gathering and management challenges associated with implementing the High-end and strategic options compared with those associated with the basic option. But most of the participants found sufficient value in the additional information produced by the high-end and strategic options to merit movement in the direction of those options. When asked, "What level of planning are you going to implement?" 80 percent of the participants indicated plans to implement either the high-end option or the strategic option for distribution system assets, whereas 89 percent indicated plans to implement the high-end option or the strategic option for plant assets.

The comments from the Pittsburgh Water and Sewer Authority's Finance Director, Kelly Sikorski about the overall importance of asset management, in her appended letter of comment, illustrate why a number of the participating utilities have indicated a willingness to undertake the additional efforts to implement high-end or strategic options:

> Asset management is a structured business process intended to minimize the cost of asset ownership, improve service reliability, and prolong the life of infrastructure systems, enabling the organization to provide for current and future customers. With all the challenges facing water utilities today, asset management provides the tools necessary for the utility to manage its way through.

Factors such as the availability and complexity of system data and the number of stakeholder groups and their specific information needs strongly influence whether it is appropriate to undertake the additional data development and maintenance required to support the high end and strategic options. For example:

- Clifford Jamile, Manager and Chief Engineer for the Honolulu Board of Water Supply, concluded in his appended letter of comment that "Since we have implemented most critical-need projects identified through our previous asset management efforts, we find substantial value in the capabilities of the strategic option to help identify the next tier of renewal and replacement projects for attention. As such, in our case it is worth the additional data development costs to have the benefit of the outputs of the strategic option."
- James Spacek, Director of the Portsmouth, Virginia, Department of Public Utilities, concluded in his appended letter of comment that "The City of Portsmouth has adopted the 'basic option' approach for its utility systems.... Both the 'high-end' and 'strategic' options have compelling potential benefits, but for our situation, the cost and complexity seem to represent an effort with somewhat diminished benefit in our case."

Most Useful Tables for Management Decision-Making

The participants also rated the usefulness of the specific sample outputs shown in Chapter 3 and in the appendix for each participating utility for both management decision-making, and for making the case for appropriate R&R budgets to governing boards and stakeholder groups. For management decision-making, Table S-1, the platform that enables multiyear capital expenditure and R&R planning across individual assets, was deemed the most useful output. This table allows the user to renew or replace combinations of assets (e.g., all assets with performance ratings below an identified score) and to fund designated percentages of the target renewal and rehabilitation funding in each year. For plant assets, this table was rated a score of 4.4 out of 5, with 5 representing very valuable and 1 representing not very valuable. This table was also rated the most valuable output table for distribution system assets, with a score of 3.9 on the five-point scale. For plant assets, all of the tables received ratings of at least 3.0, which represents somewhat valuable. For distribution system assets, a few of the tables received average ratings lower than 3.0, including Table B-3 (Replacement Cost and Asset Value) and Table B-2 (Target R&R).

Most Useful Graphics for Management Decision-Making

Of the technical study graphics, Figure S-1, the strategic option figure that shows the impact of renewal and replacement strategies on system value, was rated the most useful. This figure received a rating of 3.7 on a five-point scale for both plant and distribution system assets. For plant assets, all of the sample figures received a rating of at least 3.0 on the five-point scale regarding value for management decision-making. For distribution system assets, all of the sample figures received ratings of at least 3.0, except for Figure B-2 (Maintenance Activity), which received an average rating of 2.6.

Several of the sample outreach graphics also were rated highly as aids to management decision-making. Figure O-4 (Projected System Value) received a rating of 4.0 for plant assets and 3.9 for distribution system assets. Figure O-3 (Asset Condition Rating) also received high scores as an aid to management decision-making.

Most Useful Tables for Stakeholder Outreach

Most of the tables developed to support the asset management options are intended primarily to support the data development and management decision-making process. In most cases, the tables are more detailed than the information that would be expected to be used in making the case for support to decision-making bodies and stakeholder groups. It is therefore not surprising that most of the tables received fairly low ratings vis-à-vis their usefulness for stakeholder outreach. All but one of the tables were rated lower than 3.0 (somewhat valuable) for this purpose. The only table rated higher than 3.0 for stakeholder outreach is Table B-1 (Service Life Used).

Most Useful Graphics for Stakeholder Outreach

Four examples of outreach graphics were developed to illustrate the kinds of outputs that might be useful in communicating the results of the asset management options to stakeholder groups. These four graphics were not intended to represent the diverse range of outputs that might be needed, given different stakeholder and decision-making environments. Of the four sample outreach graphics, Figure O-4 (Projected System Value) was deemed the most valuable for stakeholder outreach for both plant and distribution system assets. This figure received an average rating of 4.3 for distribution system assets and 4.2 for plant assets. Figure O-3 (Asset Condition Rating) also received strong marks, with an average rating of 4.0 for both plant and distribution system assets. All of the outreach graphics received ratings of at least 3.0 in terms of usefulness for stakeholder outreach.

Several of the more-technical graphics also received fairly high ratings for potential usefulness in making the case for appropriate R&R investments to stakeholder groups. While these figures may be considered too complex for communication with decision boards and stakeholder groups in some contexts, the relatively high ratings for some of these figures reflects that fact that some utility participants indicated that their governing boards and stakeholder groups would expect fairly specific technical information to support requests for R&R

expenditures. The highest-rated figures of the technical study graphics for stakeholder outreach purposes are Figure S-1 (Net System Value) and Figure B-1 (Service Life Analysis), which received ratings of 3.8 or 3.9 on a five-point scale in terms of value for stakeholder outreach.

STRATEGIC USE OF DATA HIERARCHIES

The analyses conducted during this study bore out the principle that attention to strategic development of data hierarchies can add substantial value to asset management programs. In his appended letter of comment, Alan Zeisbrich, Senior Project Manager at the Santa Clara Valley Water District, indicates:

> Without at doubt I would say the asset hierarchy is the most critical aspect as it establishes the foundation of any asset management program. Much of what an organization hopes to achieve through an AM business model depends on the thoughtfulness and care designed into the reporting tool hierarchy as it dictates the type and detail of the information the system will provide.

There are several specific benefits that were found to accrue through the development of optimized hierarchies, including:

- **Time Savings**—Several of the participating utilities initially developed asset hierarchies that coincided with the level of detail of information consistent with certain existing databases, such as GIS systems, CMMS systems, and fixed asset records. For example, the databases associated with GIS systems in some cases have length, diameter, and installation dates for very small segments of pipe that are only several feet in length. As the data development efforts for the project continued, a number of these participants found that it was very difficult to populate the performance measure information meaningfully for pipe segments of such small lengths. Where it was possible to populate the performance information under such circumstances, some of these participants found the effort to be very time-consuming. When the first round of preliminary outputs for the asset management options were presented to the utility participants midway through the project, many of the utility participants realized that there was limited value for asset management purposes in descending to the lowest-level assets in some of these data systems (GIS, CMMS, others) that were constructed to accomplish other purposes. In their scorecard ranking of the options and letters of comment, a number of the participants indicated that they anticipate future time savings in building out their asset management program by building asset hierarchies for asset management purposes that start at a somewhat higher level of organization than the level at which they initiated their efforts for the sample data sets used in this project.
- **Increased Value through Strategic Development**—As discussed more fully in Chapter 6 (Appropriate Data Hierarchies), there are a number of strategies that can be used in grouping assets in an asset management system (e.g., by location, by assets of similar function and size). The experience of the participating utilities demonstrates that there is no single approach that fits the needs of all water utility systems. The experience of the participating utilities showed that the selection of an asset hierarchy

should be guided by the planning, budgeting, and financing context of the system, and by how decision-makers and stakeholder groups prefer to see information organized in order to support the allocation of resources for renewal and replacement programs. For example, in some cases, it may make sense to group all assets of a certain type and size into CIP projects, regardless of where they are located (e.g., a valve replacement program that allocates funds throughout the system), while in other cases it may be advisable to group assets by geography (e.g., replacement of pipes of a certain age or performance history within a specified neighborhood or groups of neighborhoods). Based on input from the participating utilities, the selection of a strategy is system-specific and, in some cases, asset-type-specific. Several of the participants tested different strategies for portions of their data sets or revised their approach to asset organization midway through the project. Based on their evaluations and reviews, there is a strong sense that selecting an asset hierarchy that fits the budget, CIP, and stakeholder contexts results in substantial additional value for the asset management systems.

UTILITY WORKFORCE CHANGES INCREASE THE CHALLENGES IN IMPLEMENTING EFFECTIVE ASSET MANAGEMENT PROGRAMS

There has been much written in recent years on the subject of succession planning and the effect of a changing workforce on the overall operation of water utilities. During the workshop sessions on stakeholder and asset management planning conducted for this project, many of the utility participants identified repeated reorganizations, staff cutbacks, budget pressures, organizational chaos, and lack of succession planning as posing specific challenges in pursuing consistent progress in asset management.

In some cases, these staff transitions have meant the loss of credible spokespersons to the governing boards of the utilities, resulting in reduced financial support for the programs. In other cases, the transitions resulted in the loss of staff members with unique information about the systems that had not been passed on to others within the organization or into databases and data systems, slowing progress in staff development of recommendations. The anticipated retirement of many more senior utility staff members during the next 5 to 10 years, as the Baby-Boomer cohort ages, will pose additional challenges in this area, further increasing the importance of capturing and institutionalizing the insights and information of these senior staff members during the next several years while there is still an opportunity to do so.

THERE ARE MANY RELEVANT STAKEHOLDER GROUPS, MANY WITH SPECIALIZED COMMUNICATION NEEDS

During the past decade, managers of water and wastewater utilities have become increasingly aware of the need to communicate the necessity for renewal and replacement funding more effectively to their governing boards and stakeholder groups. Previous AwwaRF studies have helped to develop tables and graphics that have become part of these communications. For example, the "Nessie curves" developed in studies for both AwwaRF and AWWA and similar graphics have been used by some of the participating utilities to communicate the high-level impact of aging infrastructure on the funding requirements for their systems. The "Supplemental Participant Studies" section of the appendix contains an example of a memorandum in which the City of Toronto uses these graphics to communicate the magnitude of future funding needs.

As detailed earlier in this section, the participating utilities were asked to evaluate the usefulness of the asset management options and of the specific sample outputs in communicating their systems' R&R needs to decision-making boards and stakeholder groups. To aid these evaluations, facilitated discussions at project workshop meetings were used to identify the specific stakeholder groups with an interest in asset management issues, and their specific information and communication needs. Many of the stakeholders identified by the utility participants are the "traditional" groups that have been mentioned frequently in recent years as the target groups whose support is needed to secure adequate funding for renewal and replacement programs. However, there were a number of surprises among the stakeholders identified—groups whose support is important in some situations but that are not necessarily thought of as target audiences for asset management information. The group discussions also produced some interesting insights regarding the different needs for information and preferred communication media for some stakeholder groups.

Several of the letters of comment address this issue. For example, in her appended letter of comment, Kelly Sikorski, Finance Director of the Pittsburgh Water and Sewer Authority, indicates that

> the Authority, like all others, has various groups of stakeholders that can benefit from an asset management program. Direct management of Operations and Engineering, senior management, the Board of Directors, regulators, city politicians, the financial community, and customers can all be benefactors of an asset management program. The challenge is to generate the necessary information in formats that are useful for each distinct group.

In his appended letter of comment, David Evonuk, Project Manager at the Portland Bureau of Water, indicates that:

> it is also important that the output tables, figures, and graphics be formatted to target various audiences including managers who prioritize which assets to spend money on, administrators and governing boards that choose which programs get funded, and the public to show that money is spent responsibly. In general it seems that detailed tables and figures are important in deciding how to allocate resources to assets, and that comparative figures and graphics are useful to gain support from governing boards and the public.

The remainder of this section identifies candidate stakeholder groups that were identified as important by the utility participants and some of the specific information and communication needs for these groups. The specific group of stakeholders whose support is critical in securing funding is context-specific. For example, in states where rate increases are directly regulated by public utilities commissions, preparing documentation of the R&R needs in a format in which the commission can support and approve the required revenues for the R&R program is likely a critical element of the communications program. In states where there is little or no direct regulation of rates, state agencies may not be significant stakeholders. Also, the best way to communicate with stakeholders is in a manner that is context-specific, depending on such factors as the size of the system and the access stakeholders have to emerging communication media such as Websites and e-mail lists. However, based on insights gained by the participating utilities through discussions with their peers at the project workshops, readers of this report who are responsible for communicating the need for R&R funding may gain insights for targeting their communication programs by considering the range of stakeholders and communication considerations presented below. In addition, a number of the participating utilities addressed their specific stakeholder groups and related communication needs in their letters of comment, which are included in the appendix.

Stakeholder Groups Identified

The following list identifies stakeholder groups whose support was identified as important by at least one of the 11 participating utilities:

1) Governing Boards
2) Municipal Government, State and Provincial Leadership
   a) Municipal Department Leadership and City Managers
   b) Mayors and City Councilors
3) The Financial Community
   a) Bondholders
   b) Underwriters
   c) Rating Agencies
4) Customers of the Utility System/the Media
   a) Retail
   b) Wholesale
   c) Citizen Advisory Committees
5) Service Providers
6) Utility Employees
   a) Executive Management
   b) Operations Managers
   c) Trade Supervisors
   d) Maintenance Managers and Business Process Managers
   e) Financial Planners
   f) Risk Managers
   g) General Employees

Communication Needs of Specific Stakeholder Groups

Utility Governing Boards

Several of the utility participants indicated that outputs such as those developed for the three asset management options studied provide the framework for increasing the understanding by water commissioners and other decision-makers of the need for increased attention to asset renewal and replacement. The governing boards' needs for information varied considerably among the utilities. In some cases, detailed information was required to support implementation of both asset management data development programs and the actual R&R activities; in other cases, a strong reputation of trust, based on historical performance and carefully reasoned requests, resulted in limited need for detailed supporting information.

Municipal Government, State and Provincial Leadership

Several of the utility participants indicated that municipal, state, and provincial leadership can be important stakeholders to asset management programs. As detailed below, there may be special information and communication needs associated with these groups:

Municipal Departments Leadership (above Utility Department Managers) and City Managers—These groups need to understand the time and resources required to implement water system asset management. There are sometimes unrealistic expectations regarding the availability of information, because other sectors (e.g., transportation) may have established reporting and investment prioritization capabilities based on years of prior data collection and investment in decision-support systems. In some cases, senior management needs better appreciation of the complexity and detail of information required to support appropriate asset management decision-making. Key messages for this group need to include demonstration of the data gaps and reasons that decision-support resources are required, along with examples that illustrate the enhanced decision-making that will be possible with the additional information. Several of the utility participants pointed out that there is an intrinsic benefit in objective, consistent approaches to help justify investments and potential impacts on rates. However, imperatives of citywide programs impacting infrastructure, such as street paving, may take priority, thus highlighting the need to collaborate and integrate asset management work across all impacted systems.

Mayors and City Councilors—City councilors need better information on the magnitude of the renewal and replacement needs, and the problems that will occur if investments are not made. They need to understand that asset management provides the opportunity to achieve long-term financial stability and to avoid major headline events that occur when major assets fail. They need to understand impacts and benefits that accrue to their constituencies and neighborhoods, and also the impacts of a sound asset management program for the system as a whole. Key information needs for this group include outputs that convey what the needs are, associated service impacts, and rate implications. Another need is to show how asset management programs can help with financial reporting and compliance needs. For example, recently enacted Canadian legislation, the Sustainable Infrastructure Act, is somewhat like GASB 34 yet likely more stringent. It will require stamped and signed Renewal and Financing Plans. Helping councilors and local government leadership understand that asset management may be a vehicle for the development of requisite plans and compliance with such regulations provides added incentive for these stakeholders to support sound programs. Some of the studied performance measures deemed to be of likely interest to this group include age as a percent of service life, number of main breaks, unplanned interruptions, condition rating, and projected system value.

The Financial Community

The financial community (bondholders, underwriters, rating agencies) needs to see a demonstration that that utilities have a plan to effectively manage and maintain the major infrastructure investments to be financed (and thereby their bond ratings). Some of the benefits of asset management programs that need to be communicated to this group include proper maintenance, regulatory compliance, management of funding demand over time, management of major risks, and not deferring costs. Key messages for this group include information that asset management will help ensure appropriate ongoing maintenance of built infrastructure, provide for sustainable facilities that can be supported with reasonable rates. Performance measures deemed to be of interest to this stakeholder group include age as a percent of service life, number of main breaks, unplanned interruptions, condition rating, projected system value, violation of internal standards, planned vs. reactive maintenance, and scheduled vs. completed maintenance.

Customers of the System/the Media

The project team identified several customer groups with whose support is important to the success of asset management programs, with sometimes varying information and communication needs:

Retail Customers—Need to be convinced of need to invest resources in proactive asset management, particularly in those situations where there are no major service problems (for example, relatively new systems). Some of the benefits of greatest interest to customers include low and stable costs and rates, reliable service, including quick response to service outages. Key Messages for this group include the information that shows that it is cheaper in the long term to conduct proactive renewal and replacement, and that even relatively new systems need to do asset management to preserve high quality services over time. Some of the performance measures deemed to be of greatest interest to customers include violations of internal standards and unplanned interruptions in service.

Wholesale Customers—Need to accept the level of unmet needs and prospective capital requirements into the rate base or other rate formulas that they will be paying. The key message for this group is that asset management provides a vehicle to identify, prioritize and report back on investments, and a vehicle for identifying investments that lead to overall, long-term efficiencies and supportable rates.

Service Providers

Service providers, such as engineers and construction managers who help to implement the identified and adopted renewal and replacement programs, can be important stakeholders to timely delivery and execution of the R&R programs. Some specific data needs of this stakeholder group were identified by the utility participants.

This group needs information on where the capital improvement program is headed and scheduling requirements to be prepared to support preliminary engineering and ensure ability to affect timely delivery. A key message for this group is that asset management programs provide timely, systematic information on prospective direction and focus of capital investments.

Employees of the Utility

Several groups within a utility organization have been identified as important stakeholders with special communication needs.

Executive Managers—Need to manage expectations regarding time required for implementation of asset management and potential cost savings. Also needs justification for why, in spite of competing programs and limited resources, asset management should not be deferred until completion of strategic planning, CIP planning, and/or master planning projects. Key messages for this group include realistic information regarding the timing and resources required to collect detailed data over a wide array of assets, and that the likely outcome of analysis will not be a near-term determination of lower investment requirements. Rather, asset management will enable better distribution of investment to achieve greater benefit in the long-term. Another key message is that asset management may be done in parallel with other utility planning efforts and, in fact, will help improve the quality of these undertakings. Asset management will provide significant benefits in improving prioritization of R&R spending within the utility's CIP and may help avoid having capital spending be used to address O&M problems.

Operations Managers—Need to understand that outputs of asset management will help them manage their departments and divisions more effectively and efficiently. Key information needs include demonstrating that Operations personnel are already collecting 90 percent of required information and limited additional effort, combined with improved organization and management of the data, will yield substantial returns.

Trade Supervisors—Need to be able to effect greater control of work order dispatching to limit the sometimes helter skelter nature of current activities. Key messages for this group include illustrating how asset management can enable more effective approaches to identify and plan work order assignments, prioritize investments.

Maintenance Managers and Business Process Managers (Distribution, Treatment, Transmission System Managers)—Need to understand that outputs of asset management will enable them to perform responsibilities more efficiently and effectively. Key messages for this group include the fact that staff is often already collecting a lot of the information required to support asset management programs, but it is currently not stored and used in a manner that takes maximum advantage of the information. This group needs to be shown that the additional data collection that is required and systematic use of information will not be burdensome, and they need to see how the outputs created through the more advanced options can be useful in guiding their maintenance planning and in securing funding support for adequate levels of maintenance activity.

Financial Planners—Need to get better, more detailed information on R&R needs of utility systems and the associated rate implications. Key information needs for this group include demonstration of how asset management provides analytical support for R&R investment and helps prioritize across competing demands for limited resources.

Risk Managers—Need to understand how the prevailing risks in the system and utility's approach to risk mitigation or management are influenced by aging infrastructure and R&R policies. Key information needs for this group include demonstrations of how including risk assessment as part of the asset management program supports the overall risk management assessment needs of the system. Risk managers and staff responsible for developing and implementing asset management programs have mutual need for collaboration on data and strategies.

General Employees—Sometimes ignored as a stakeholder group, utility employees can be important points of outreach to their neighbors, relatives, and others in the community. Developing targeted communications through employee Websites, e-mail communications, and newsletters should not be ignored as an overall communication program is developed.

Elements of the Message

Based on the evaluations of the utility participants, the sample figures and tables developed as part of this research project can form an important part of the stakeholder communication messages related to asset management. Other elements of the message that were identified by the project team over the course of the project include the following:

- Pictures of failing assets or which illustrate asset condition
- Pipe samples (at public meetings)
- "Nessie" curves
- Statements of executive level policy guidance and goals
- Illustrations of levels of service options, and facilitated discussions to identify targets for a system
- Articles and other communications that publicize benefits to community of action (e.g., better service, support to economic development objectives, sustainability of the system)
- Return-on-investment analysis (benefit/cost)
- GIS (or just plain maps) showing condition, age, risk, possibly with links to hydraulic model

CHAPTER 10
RECOMMENDATIONS

This chapter identifies recommendations that emerge from the technical studies and evaluations conducted as part of this project. Overall, the study team found the use of prioritized performance measures to be a useful and valuable way to characterize assets and guide renewal and replacement investments. The recommendations fall into three categories:

- Applicability of options and outputs studied to water utilities
- Recommendations when implementing the options and measures studied
- Suggested future research

APPLICABILITY OF OPTIONS AND OUTPUTS STUDIED TO WATER UTILITIES

As detailed in the Chapter 9, the participating utilities found the asset management options and performance measures studied to be valuable. Many of the participants found value in the basic option, and most of the participants indicated that the additional value added warrants movement toward the high-end option or the strategic option. Eighty percent of the participants plan to implement a program along the lines of the high-end or strategic options for distribution system assets, and 89 percent plan to implement a program along these lines for their plant assets. As detailed in Chapter 9 (Findings), many of the performance measures studied were rated 4.0 or higher on a five-point scale (with 5.0 being best) in terms of value added to either management decision-making or stakeholder outreach. During workshop discussions, the participants indicated that they saw specific value in having the information tested in making choices on how to allocate scarce funds available for renewal and replacement programs, in determining the overall level of funding, and in making the case to specific managers, board members, and stakeholder groups who are interested parties to their utilities operations.

While the evaluations identified strong value in moving toward the options studied, the data-gathering field test conducted for this project confirmed that there are considerable challenges in gathering and verifying consistent data for even the basic option studied. The data needs and challenges associated with the high-end option and the strategic option are even greater. The participants on balance deemed the additional investment required to implement the high-end and strategic options, which are based on performance measures, as cost-beneficial in light of additional insights and value gained.

A number of strategies were identified to improve the likelihood that the options studied during this project could actually be implemented, given the real-world constraints of budget limits and staff resource limits. The following section details the specific strategies that were identified. As an overall approach that transcends specific strategies and tactics, a utility could start with a lower level option, such as the Basic Option, and build upon that as resources become available. A number of the participating utilities, such as the Honolulu Board of Water Supply, have indicated plans to follow exactly that overall approach to build upon their existing asset management work. With careful planning related to data formats and platforms, the data developed to support a basic asset management program can be efficiently ported support a higher-level program as needs warrant and as resources become available.

RECOMMENDATIONS WHEN IMPLEMENTING STUDIED OPTIONS

Budget limitations and competing demands for staff resource time preclude complete, immediate implementation of the asset management options studied during this project. Based on the experience of the project team in conducting this project, and the experience of the participating utilities and other members of the project team in other related endeavors, a number of strategies have been identified to enable proactive, real-time development of meaningful asset management databases and decision-support systems that add substantial value in during the early years of implementation.

The value of these strategies is context-specific. Not all of these strategies are appropriate for all water utilities. In some cases, one of these strategies can be implemented as the overall framework or approach for implementing an asset management program. In other cases, several of these strategies might be combined to define a viable, implementable program. Many of the letters of comment developed by the participating utilities, and included in appendix to this report, illustrate the use of the strategies described below. The strategies are grouped into categories, depending on whether they primarily relate to the following:

- Asset selection/data management
- Data management
- Use of data hierarchies
- Use of performance measures

Asset Selection /Data Management

Address Most-Critical Systems First

To make asset records consistent and data reliable takes time but is worthwhile in the long run. Because of this, it is important to focus first on the most-critical assets and types of information. Resource limitations, plus the disruption to other activities that would result from data gathering and verification processes, make it difficult to fully implement asset management programs such as those studied in this project systemwide within a very short timeframe. Gathering data first for assets deemed most critical or in need of likely R&R interventions is a reasonable point of departure. Where utilities have conducted criticality studies as part of a vulnerability study or other planning effort, these results can help to guide the selection of "first-in" assets. Even where such formal assessments have not been conducted, staff judgment can be used to create a logical plan that provides more value than random introduction of assets into the databases. In selecting a small portion of their systems to study for this project, several of the participants were able to identify portions of their plant and distribution system assets for which the findings of this study could be used to aid real-time decisions regarding which of several competing sets of asset should receive priority funding for renewal and replacement activities.

Implement First the Processes to Capture New/Replaced Assets

For utilities that have limited, inaccurate, or otherwise difficult-to-gather legacy information about existing assets, a high initial priority is to establish and immediately implement revised data-gathering protocols to capture the kinds of information deemed valuable for new assets added to the system or for replacement assets brought into the system as existing assets are exhausted. Early implementation of data gathering for new or replaced assets provides an early data set that can be used to validate the data entry and reporting for the asset management system, even while difficult-to-gather-and-validate information for older assets is being developed. Another benefit of this early implementation of rules for new assets, particularly in rapidly growing systems or ones with aggressive ongoing renewal and replacement programs, is that it prevents the growth of an ever-expanded backlog of assets to be "fixed" for entry into the asset management system.

Recognize That There Are Often Substantial Gaps

Even for assets that are thought to be fairly complete and up to date for financial and condition/performance data, there may be a surprising number of either blank records, incomplete information, or information that is found to be invalid when tested. A number of the participating utilities for this project began the data collection phase with the expectation that some of the performance or description fields would be easy to populate electronically from existing systems, only to find that a considerable effort was need to backfill or clean up the information for these measures.

Financial valuation and maintenance cost data for many water systems are incomplete and are often available only at a very high level. Some of the participating utilities, such as the San Francisco Public Utilities Commission and the Santa Clara Valley Water District, went through a manual allocation process in order to assign high-level cost estimates to individual assets to populate the financial and related performance fields for this project. Even when such records were based on estimated values, this information was found in some cases to provide very useful insights regarding where investments are being made, and the recurring maintenance costs associated with certain types of assets. In his appended letter of comment, Alan Zeisbrich, Senior Project Manager at Santa Clara Valley Water District, points out:

> Data availability and accuracy was a big concern of mine because I know we did not have good replacement cost data or complete equipment attribute information....What is important when considering data availability is reasonable and consistent application of estimates and methodology where the cost data is not discretely available. Also, keep in mind the importance of momentum and "just get started."

By reviewing the groupings of maintenance expenditures into the levels of hierarchy developed for this project, some of the participating utilities gained insights regarding the differential level of investment and expenditures on some of their distribution system elements that proved useful in redirecting the expenditures to achieve a greater balance of attention to the varying systems.

Another challenge in populating the financial information for many of the participating utilities was that most utilities do not have a mechanism to efficiently associate the soft, nonconstruction costs associated with installing specific assets to those assets. As a result, the full investment associated with replacing or rehabilitating those assets is often understated. Even if this information is available only at a higher level of organization, it is important to capture it and consider it when making renewal and replacement plans for assets, as these costs can be substantial, and stakeholders and decision-makers will not be adequately informed to make sound decisions if these costs are ignored.

Explore Options for Electronic Data Conversions

It pays to explore options for electronically converting and addressing asset inconsistencies before falling back to manual fixes. While not always successful or possible, the electronic conversions and fixes are much more efficient when they are available and work. In addressing the data clean-up required for the sample data sets developed for this project, several participating utilities were able to develop adjustment routines that filled in or corrected information that was incomplete or inaccurate, by exploring options to regroup information within other database or to combine information from multiple separate systems. This allowed them to avoid the very time-consuming manual cleanup that some other participants had to perform, because they did not have the option to use electronic options. The ability to electronically backfill will depend on such factors as the programming platform for existing electronic information, the compatibility of multiple external databases, and reasons for the incomplete or inconsistent information. But, given the time savings and resource savings that can be realized where possible, it is worth carefully exploring electronic conversion/correction options before initiating laborious manual fixes.

Conduct Beta Tests of the Data-Gathering Effort to Validate Both Information and the Data Approach

Even within the confines of developing a 1,000-record data sample for analysis for this project, several of the participating utilities found that there was substantial value in field-verifying the data for a couple of dozen records before launching into full production development of the 1,000 record datasets. In several cases, this exercise helped to ensure that data collection was complete, and allowed development of work-around strategies that would have otherwise required utility staff members to go back later to backfill or correct information.

Be Aware of Confidentiality Considerations When Mapping Out a Data-Gathering and Integration Strategy

Many water utilities have a heightened appreciation of confidentiality considerations related to their systems in light of the heightened focus on security during the past several years and through conducting vulnerability assessments. Some of the information that is required to best prioritize renewal and replacement needs may be sensitive for a variety of reasons. In addition to security considerations, some information may relate to legally sensitive information that affects contractual relationships with customer communities or individual customers. When developing the overall information needs and usage plans for the asset management data registry, these confidentiality considerations should be considered.

Hierarchy

Understand and Define Your Asset Management Units

A significant lesson from this project is that the definition of an asset management unit in the asset registry, and the linkages that developed to report upward to higher level summaries, has a critical role in the success and usefulness of the information in aiding both management decision-making and communications with decision-making bodies and external stakeholders. The definition of what constitutes the lowest level of asset in the system is significant decision, and should include input and review from all of the groups within the utility who are likely users of the information being developed (engineering, finance, planning, directors, information technology groups, and others, as appropriate).

Start at a High Level

Where detailed information on individual assets is limited, incomplete, or otherwise challenging to gather in the near-term, one useful strategy may be to enter information at a higher-than-planned level of hierarchy, and then to descend strategically down to further levels of detail as resources and staff time permit. Having even reasonable high-level estimates of asset value, age, and performance is often very valuable and in many situations worth populating as an interim aid to decision-making while more detailed information is being developed.

Be Wary of Just Adopting Detailed Information because it Is Available

When considering how to build the asset register and populate it with descriptive and performance information, it is important to draw upon the resources of information already available in related electronic databases, such as GIS, CMMS, and financial/accounting fixed asset databases. However, there is a potential substantial pitfall in incorporating information from these systems directly into the asset management program without review and reflection. The most detailed asset level in CMMS/GIS systems is *not* always necessary for R&R planning. These systems may contain information at a far greater level of detail than is necessary for determining renewal and replacement needs. For example, GIS systems may have pipe segments broken down into records as small as 3- or 4-ft sections in some areas. Loading these records into an asset management system can create multiple problems, such as:

- Maintaining the basic descriptive information for assets at the level of 3- or 4-ft segments of pipe can add a substantial database maintenance burden in the long run
- Gathering and storing performance data at this level of detail can be difficult, and sometimes impossible. Assigning service outages to discrete several-foot segments of pipe, for example, could be a laborious and ultimately not very useful exercise

Several of the participating utilities began their datasets for the project by extracting very detailed records from GIS or CMMS systems. They found the effort of populating financial and performance measures at this detailed level to be burdensome, and in some cases without substantial meaning. Ultimately, when they saw the graphic and tabular results from the template model at this level of detailed, they decided that there was no need to maintain asset information at this level of detail for the purpose of identifying and prioritizing renewal and replacement investments, which would be defined at a higher level. As detailed several of the participating-utility letters contained in the appendix, several of the participants decided to aggregate some of these assets at a higher level of organization between the draft and final rounds of data development for the project. Several of the participants who elected to maintain their project datasets at a very detailed level have indicated that they would likely aggregate assets at a higher level of hierarchy for future work in asset management support systems.

Plan the Data Hierarchy around CIP and Other Funding Processes

One strategy for determining an appropriate level of detail for the data hierarchy is to start from the CIP planning process, the budget process, and any related financial-planning processes. There is no single answer to this issue because water utilities' governing boards and stakeholder groups vary considerably in the level of detail at which they approve projects for action. While size of the utility can be a contributing factor, the decision-making culture and enabling state and local legislation can be equally significant in shaping the level of detail at which financial decisions are made. By working backwards from the level of detail at which these financial decisions are made, the utility staff can identify the level at which asset condition, age, and other information need to be organized to support the CIP and budget processes. Additional levels of detail can be added later, as time and resources permit, to support additional analyses that staff deems important to determining their recommendations to external stakeholders.

Consider Using the Hierarchies of Other Related Databases

To facilitate electronic exchange of information, consider using the hierarchy (or at least the asset identification numbers) from other related databases within the utility, such as CMMS, GIS, and fixed asset systems. While the decision may be made to deviate from the hierarchies used in these systems because the needs for asset management planning are different or more specialized, the hierarchies within these other databases should at least be considered for adoption. Even if the asset management system is developed and maintained at a more summary level of organization, it may be possible to use the asset identifiers, coding, and description at that higher level from the related databases, to facilitate sharing of information and use of the outputs of the several systems.

Performance Measure Information

Justify Strategically the Addition of Each Performance Measure

Participants' reviews of outputs for the high-end and strategic options confirmed the strong value of a performance measure–based approach to identifying assets that are candidates for renewal and replacement. The evaluations also demonstrated that there is no standard set of performance measures that applies to all utilities. While some of the measures were rated highly by most of the participants, some saw strong value in measures that were of limited interest to other participants.

Given the substantial additional costs to gather, input, and maintain each new performance measure, particularly when applied to the whole population of thousands of assets, each performance measure should be carefully evaluated in terms of strategic value to either management decision-making or stakeholder communications before being added to a utility's asset management system and strongly justified on a benefit-cost basis. Utilities should resist the tendency to add some measures because they might be useful, or because the data happen to be readily available for some portions of the asset base. In his appended letter of comment, Alan Zeisbrich, Senior Project Manager at the Santa Clara Valley Water District indicates that

> …if the measures are complicated or burdensome, they are probably not good measures by which to evaluate your performance. My advice is simple: develop only a handful of measures, make them meaningful, and track them religiously. What I mean by useful is also quite simple. Ask yourself: what do I really need to know to manage these assets? Don't get caught up measuring things that don't provide objective quality evidence that can and will take action on.

This is especially true because the information required to populate a single performance measure may come from multiple sources. For example, when determining a strategy to collect information for the "service outages" measure used in this project, the San Francisco Public Utilities Commission identified six potential data sources (CMMS system tracking information, main break reports, leak detection analyses, manually collected information on unaccounted for water, water supply information, and information contained in GIS databases), and the Honolulu Board of Water Supply identified three sources of information (information from HBWS's work order system, GIS data, and equipment failure–tracking systems).

When identifying performance measures to include in a utility's asset management program, useful resources include:

- Measures actually used in this project (detailed in Chapter 2)
- List of candidate performance measures identified for consideration at the start of this project (also included in Chapter 2)
- Performance measures identified in the AwwaRF study "Selection and Definition of Performance Indicators for Water and Wastewater Utilities," which was released by the Research Foundation and AWWA in 2004

Also, when establishing a new or revised framework for asset management planning, it is worthwhile to consider whether any performance measures currently being tracked can be discarded. That is, just because certain information has been gathered and processed historically does not mean that it has value in light of changing approaches and agency objectives. Just because historical information is available for a given measure does not mean that scarce future resources should continue to be devoted to developing updates to it.

Recognize That Some Performance Measures May Need to Be Tracked and Addressed at Different Levels

Depending on the specific performance measures that a utility deems worth capturing and using as part of its asset management decision-support systems, it may be necessary to address some measures at different levels of hierarchy. The databases and decision and planning frameworks should be adapted to accommodate such variations, where necessary. For example, when evaluating the value of the performance measures studied as part of this project, many of the participating utilities concluded that some of the measures, such as violations, were difficult to assign to individual pipe segments, and probably should not be. Other measures, such as number of unplanned service interruptions and R&R status (whether an asset had been relined, rebuilt, or protected), are meaningful at the asset level and ought to be tracked at that level. Developing a decision-support system that can accommodate such diverse measures at their appropriate levels of hierarchy is important to providing efficient and meaningful information about assets to guide decision-making processes. Also, some participants identified a variation between plant and pipe assets in terms of the level of organization at which it was appropriate to capture and use performance information.

Weighting of Performance Measures Is Important

Customizing the weights assigned to performance measures was found to strengthen the alignment between the decision-support system and a utility's core values. More than half of the participating utilities used the option that was provided to customize the weightings from a default set of weightings that was developed by the research team on the basis of discussions with the overall project team. In some cases, such as with Portland, Oregon, and the Pittsburgh Water and Sewer Authority, this was done to reflect the fact that data for some of the performance measures were not available or were deemed of limited reliability. In other cases, this was done to specifically reflect the fact that some of the performance measures were viewed as strongly in alignment with agency goals, while others were less important than the default priorities would suggest.

Customizing the Scaling of Performance Measures Adds Value

As detailed in the Chapter 2 (Methods), the template model developed to test the three asset management options included a default scaling factor for each of the performance measures studied. But, the model allowed the opportunity to customize the scaling of the performance measures used in the high-end and strategic options. For example, for the "Service Reliability—Interruptions" performance measure, the default value deducted three points on a 10-point scale for every two service interruptions recorded during the preceding 2-year period. The Santa Clara Valley Water District customized the scaling for this measure to deduct three points for every single service interruption. In this case, the customization resulted in a stronger measurement of interruptions compared to the default. For example, with the default scoring, an asset that has only one interruption would still score a perfect "10" for this measure.

Customization of the scaling, when combined with the weighting of the performance measures on a 100-point scale, was found to substantially increase the ability of decision support systems for the high-end and strategic options to identify assets whose renewal and replacement would most strongly help to accomplish water utilities' goals and objectives.

Integration with Other Agency Activities

Look for Opportunities to Integrate with Other Systems and Processes

Regardless of whether the data are stored in the exact same database or system, it is valuable to coordinate the development and use of asset management decision-support systems with other related processes and systems within the utility. This includes GIS and CMMS databases but extends beyond them to information developed and used in generating a utility's strategic plans, audit reports, and other management-support systems. For example, in his appended letter of comment, Michael D'Andrea, Manager of Infrastructure Asset Management for Toronto, identifies several points of useful integration:

> The City has now scoped the needs for an asset management system which is expected to be integrated with the work management systems….A first priority is the implementation of a system for the sewer and watermain assets. The system is expected to draw on road reconstruction renewal needs identified by the Transportation Services Division and from condition assessments of sewer and watermain infrastructure in the Toronto Infrastructure Asset Data Standard (TIADS) data registry to establish longer term strategic direction for capital program planning and assessing funding requirements, provide discrete (at pipe segment level) analysis of renewal needs, and assist with priority setting and capital program planning on an annual basis.

American Appraisal Associates, as part of its contribution to this project, identified strong potential value of the information developed for the asset management options studied in this project to the finance departments, in the aiding the development of more accurate and efficiently developed auditable records of the utility system assets. The following excerpt from AAA's letter, which is included in full in the appendix, identifies some of the key benefits to the finance departments:

> A lot of detailed information is collected and maintained at the plant level that can also be used by personnel in the Finance Department. This would require the "Asset Management" detail to be rolled up into more of a summary record, or "group" record as the Finance Department does not report at the level of detail that the personnel working in the plant need for Asset Management. From experience in appraising Water Utilities we find that what appears in the capital asset ledger maintained by the Finance Department is different than what is actually in the plant. This is because there are normally no established procedures for plant personnel who are familiar with what is occurring in the plant to report retirements and additions to the Finance Department. Their primary responsibility is to manage the performance of the plant to be sure the customers are served with the highest quality service possible. As a result, the information that gets recorded in the capital asset ledger, and used for financial reporting, is usually derived from Accounts Payable (additions) and entered as "group" entries (total project cost) in the capital asset ledger rather than as individual capital assets.

During appraisals of water utilities we find a significant number of assets on the record that are no longer in service or have been disposed of. This results in overstating the Utility's investment in Capital Assets. Capital assets consist of Land, Buildings and Improvements, Equipment, and Infrastructure and are reported by these accounts. The assets that are found to be no longer present are reported as unrecorded retirements. Consequently an adjustment has to be made to the value of the capital assets on the financial statement; usually as a write down. If an electronic interface could be established that would allow plant personnel to automatically report to the Finance Department assets that are being constructed, replaced or disposed of, it would improve the accuracy of the capital asset ledger, reduce the manual effort in the Finance Department to capture this information, and allow the Finance Department to report better information on the Utility's financial statement. This information could come from the Asset Management System.

The multiple benefits of reaching out to groups such as the finance departments include:

- These groups have databases or other systems that contain some of the information that will be valuable in populating the asset management system
- Sharing information across these systems provides for more consistent information being given to outside stakeholders such as regulatory agencies, external auditors, and funding agencies
- Since reports from the asset management system can aid several types of agency decision-making, sharing of effort in many contexts provides the opportunity to get substantial additional value from the asset registry, often with limited additional transactional costs
- Helping to meet the needs of other groups within the utilities provides the opportunity to generate additional support for the build-out of the asset management registry and condition assessment information. That is, when budgets for the full development and maintenance of the asset management databases are up for approval, there can be substantial value in having support from other groups within the utilities that will also derive benefits from the information. This is particularly true as utility budgeting processes are likely to feel the competitive pressure of stakeholders and governing boards interested in minimizing rate increases to customers

Look for Opportunities to Centralize Information

After considering the data-sharing and data conversion challenges encountered in developing the case studies for this project, several of the participating utilities concluded that there should be only one master data registry/database if at all possible. Data reconciliation problems arise by having duplicate information in multiple places (such as CMMS and asset management software).

Given competing needs for information and levels of detail of information needed for purposes not related to asset management planning, it may not always be possible to achieve this objective. However, it is a useful starting point when developing a program, with modifications to accommodate deviations as needed. Ultimately, several of the utility participants concluded that asset management should be viewed as an approach or policy that utilities implement, which may include implementation of may tools and pieces of software that help the utility to function. Asset management is not itself new software. Instead, it is a way of doing business, a business model that employs existing or new tools to accomplish its objectives. Business model terminology fosters discussion of principles and methodologies, rather than pointing discussion towards specific pieces of "software."

Centralization of information, if ultimately deemed feasible, is likely to occur in several steps, as illustrated in the asset management maturity model shown in Chapter 5. Several of the participating utilities describe their progression in this regard in their letters of comment. For example, in her appended letter of comment, Kelly Sikorski, Finance Director at the Pittsburgh Water and Sewer Authority points out:

> At the onset of this project, the Authority independently was in the development phase of an asset management program. With regard to technology, only a very limited Geographic Information System (GIS) was in place. Today, due to a very aggressive and committed IT implementation effort, the Authority, in addition to replacing the finance and customer service systems, has implemented three new fully integrated systems. These new systems include fixed asset accounting, project management, and computerized maintenance management....A future GIS upgrade is included as part of this comprehensive plan.

In his appended letter of comment, David Evonuk, Project Manager at the Portland Bureau of Water Works notes that:

> The major obstacles to overcome in implementing a Strategic program appear to be entering all asset data in electronic format, organizing asset data in such a way that information from various workgroups can be linked or accessed from a central point, and ensuring that data is updated and maintained over time. Implementing a strategic asset management plan will help in focusing resources on improvement projects that provide the most benefit, will assist in justifying these management decisions to administrators and the public, and provide a solid basis for determining the level of funding for system improvement programs.

SUGGESTED FUTURE RESEARCH

Conducting a research project like this provides insights to a number of questions of interest to the water community. By drilling into the detailed data files of a dozen utilities to create a casebook of sample outputs, this work also helps to identify remaining data gaps and research needs that go beyond the scope of the current effort. The list below identifies some of the key remaining research needs that have been identified by the research team and utility participants at the close of the technical work on this project:

1. Performance indicators and measures—while there has been some recent work on this topic by AwwaRF and others in addition to the work conducted as part of this project, the utility participants concluded that the water industry would benefit by having some additional work done in defining performance indicators/measures specific to asset

management needs (i.e., to extend the evaluation and testing of the measures used in this study and others).

2. Risk analysis—additional analysis of the value of adding a risk and condition assessment component to build a real asset management business model.
3. Expected lives—research into expected service life and factors that affect it for various classes of assets and some way of benchmarking vs. similar utilities within the same region. In addition, there may be value in researching the connections between service life and main break history—connecting prior AwwaRF work on this topic to the current study's findings and decision-making models.
4. Consequence and criticality analysis—additional work in establishing a rating system based on a utility organization's values and risk tolerance. Seattle Public Utilities, one of the utility participants in this study, has done considerable work in this area. Research efforts and publications to make such analyses more generally available to the water industry would provide added value in addressing this important topic. Three of the participating utilities tested some form of consequence or criticality measure as part of this study. Based on this limited testing, these measures appear to merit more detailed evaluation.
5. Research into relative importance of hydraulic performance (c-factor) and maintenance/repair/replacement cost-benefit analysis—these measures were tested by the City of Toronto as part of this project. Additional research and field examples from multiple contexts would provide additional guidance to the water industry on the usefulness of these measures which, based on testing conducted so far, appear to provide highly useful input to the R&R decision-making process.
6. Research into the disconnect between complaint-driven decisions and engineering judgment of utility staffs—there is a sense among the project team for this study that consideration of how to balance complaint response and long-term cost-effective solutions remains a topic ripe for careful scrutiny.
7. Cross-industry analysis—benefit could be derived by additional consideration of cross-over and opportunities for coordination with other public works activities, such as road rehabilitation programs. In addition, lessons learned in implementing asset management programs in other sectors, such as transportation, aviation, electric, and gas, have not been fully mined.
8. Organizational studies specific to asset management—in light of this study's findings that agency-staffing transitions and other organizational issues have a substantial influence on water utilities' ability to make progress in implementing enhanced asset management programs, there would be value in conducting research on organizational models that can be used to support development and ongoing implementation of sound asset management approach. Such research would address issues such as staff recruitment and retention for key skill areas needed for program success, dispersion and documentation programs to ensure continuity of efforts when staffing transitions take place, and governance frameworks that support the development and implementation of sound asset management programs.
9. Extensions of modeling effort and case development from the current study—follow-on efforts to develop additional case examples for additional utilities. In addition, testing the impact of scaling up the test to include more types of assets, more records, with related guidance to the water industry. Follow-up efforts to address the best ways fill

data gaps when scaling up models to include the number of records required to include the full asset base for a large or midsized utility system.

10. Additional research on tools and programs to reach out to stakeholders and decision-makers—The current study included development of several examples of outreach graphics and tables that might be used to communicate the results of the asset management options studied. As detailed in the recent AWWA publication *Avoiding Rate Shock: Making the Case for Water Rates*, the likelihood of gaining support for required rate increases improves substantially when there is an organized approach to stakeholder communication that is integrated early on with technical rate and financial planning studies. Although the Rate Shock report contains several case studies where an asset management approach has helped make the case for local funding of infrastructure R&R, more work in detailing the specifics of successful case examples and communication strategies keyed to specific asset management approach would provide useful insights beyond the current literature. For example, the likely stakeholder messages for a utility implementing a "basic" service life-based asset management program might differ substantially from the messages for a utility implementing a risk-based Strategic asset management program. A guidebook that defines such distinctions and provides sample outputs and messages keyed to these strategies would likely aid many utilities in fine-tuning their communication programs.
11. Asset management as a way of life: educating utility decision-makers—In recent years, conference agendas and workshop training programs have increased the understanding of utility managers and staffs regarding the scope of asset management, and the range of options and approaches available. The current project is part of that effort. Additional effort to develop materials and conduct training programs that educate the governing boards and other decision-makers on these issues may be critical. Such education would lay the foundation for the use of performance and risk-based tools, such as those tested in this study, so that there is a common language and understanding of core concepts as the specific recommended programs are carried forward to decision makers.
12. Potential for GIS application—Several of the participating utilities have either already used GIS outputs as part of their asset management communications programs (e.g., Portsmouth, Virginia and Honolulu Board of Water Supply). Further investigation of the opportunities to support and represent asset management conclusions through such methods could aid the more widespread use of these approaches. Some of these utilities, such as Honolulu are in the process of further refining their use of these tools, both as a way to gather and store information, and as a means of strategically displaying findings of the asset management studies.

CHAPTER 11
OVERVIEW OF APPENDIX

The body of this report describes the methods used to conduct this study and presents the findings and recommendations that resulted from the development and evaluation of participating-utility data sets for the three asset management options studied. The appendix material, which is on the accompanying CD-ROM, is described below.

PARTICIPANT EVALUATIONS AND SAMPLE DATA SETS

The appendix contains the individual contributions from the participating utilities, which are summarized in the Findings and Recommendations chapters of this report. In addition, it includes the detailed sample outputs for each participating utility's data sets. This section of the appendix is organized alphabetically by utility; within each utility's section, three elements are included:

- Scorecards
- Letters of comment
- Reporting samples

Scorecards

For each utility, except Ann Arbor, Michigan, this section contains a detailed, completed scorecard that evaluates the options and performance measures studied during the course of this project.

Letters of Comment

For each of the participating utilities, except Ann Arbor, Michigan, this section contains a letter of comment. These letters typically include a discussion of the utility's background context in asset management and comments on specific challenges and opportunities encountered during the conduct of the study. In addition, the letters contain supplemental commentary on the participants' assessments of the options, performance measures, and other elements studied.

Reporting Samples

For each of the participating utilities, this section includes samples of the tables and graphics that were developed for the basic, high-end, and strategic options studied. In most cases, there are 22 sample graphics used to illustrate the range of outputs and to illustrate the ability of some of the figures to produce comparisons at various levels of asset hierarchy. Chapter 3's graphics (for the Santa Clara Valley Water District) illustrate the outputs included for all of the utilities in the appendix. In a few cases, where the asset hierarchies employed did not support comparisons at all of the identified hierarchy levels, there are fewer than 22 sample outputs.

For two of the utility participants, two sets of reporting samples are included. Toronto, Ontario, elected to develop two sets of input data, so that complete neighborhood segments of older and newer portions of the system's distribution network could be included. Since each of the neighborhood segments approaches the 1,000-record data limit set for the template model used to conduct this study, two sets of outputs were generated and included in the project report appendix. Seattle Public Utilities included supplemental performance measures for criticality and consequences that were not available for the entire set of contributed assets. Two sets of outputs were therefore developed for Seattle. One set covers the entire data set, and other one includes only the assets for which criticality and consequence ratings were available.

SUPPLEMENTAL PARTICIPATING-UTILITY STUDIES

Two of the participating utilities contributed additional studies and materials to the project report.

Water Loss as Guide to R&R Programming

Ottawa has developed tools and approaches that allow it to use information regarding unaccounted-for water as an input to the development and prioritization of renewal and replacement projects. The other participating utilities and the project research team found Ottawa's approach to be useful and likely of interest to other utilities. Ottawa documented its approach for inclusion in the project report.

Sample Memorandum Used to Communicate R&R Needs

During the course of the project, Toronto shared with the project team a memorandum that was used to communicate the needs for renewal and replacement funding to its Public Works Committee. The other participating utilities and the research team found this memorandum to be a very useful example of how to communicate such program needs to decision-makers and other stakeholders. Toronto gave permission to include this memorandum as an example that may be of interest to readers.

Use of Customer Billing and Information System to Generate Work Orders

During the project workshops, the San Francisco Public Utilities Commission representative described how the SFPUC's customer billing and information system is used to help identify work orders for asset maintenance, renewal, and replacement. Based on input from the other participating utilities, the SFPUC approaches and tools in this area provide a capability that most water agencies do not have. SFPUC agreed to document its approach and tools, so that other interested utilities might consider the benefits of developing and implementing complaint-tracking systems with these capabilities.

OTHER RESEARCH TEAM CONTRIBUTIONS

Several items developed to support the discussions at the project workshops were deemed to merit inclusion in the appendix. These items are described below.

Relationship of Asset Management Options to Financial Reporting

Throughout the project, American Appraisal Associates has served as a resource and advisor on current and potential future connections between the asset management programs and databases maintained by utility finance departments and other groups within utilities. Paul Gruenwald, who is a Vice President of AAA and Managing Principal for the Milwaukee Property Appraisal Services Group within AAA, served as AAA's representative to this project. Mr. Gruenwald developed a letter report on this subject, which is included in the appendix. The report identifies important opportunities for greater cooperation between groups within a utility as there is movement within the industry toward more detailed and sophisticated planning for asset renewal and replacement. Opportunities to increase the accuracy of both planning and financial tracking information within utilities are identified. Key points from the letter are summarized in the Findings and Recommendations chapters of the report.

Presentation and Corporate Plan by South East Water

South East Water LLC (SEWL), an Australian utility that has been an international leader in the development and implementation of performance measures as aids to asset management planning, has served as an advisor throughout the implementation of this project. Gordon Logan, who developed SEWL's Water and Sewer Infrastructure Asset Management Manual and related water and sewerage system asset management plans, has served as SEWL's representative to the project. Mr. Logan has provided two items for inclusion in the appendix. The first is the presentation that Gordon made at the first project workshop, in which he described the tools and methods used to define and implement strategic performance measurement at SEWL. The second item is a version of SEWL's Corporate Plan, from which company-sensitive information has been purged, that illustrates the way that financial and performance measures are used as part of the utility's corporate planning process. Even with company-sensitive figures removed, the U.S. and Canadian participating utilities found the Corporate Plan to provide very useful insights into the use of performance measurements in the planning process. SEWL gave permission to include the "purged" version of the Corporate Plan in the appendix for readers' benefit.

Growing Importance of Rate Structures and Rate Setting for Water Utility Ratings

One of the issues addressed throughout the project is the increasing importance of proactive rate planning and long-term financial management in maintaining sound credit ratings for water systems' bonds and other long-term financial obligations. James Wiemken, who serves as Standard & Poor's sector leader for municipal water and sewer utility ratings and state revolving fund ratings throughout the United States, was invited to make a presentation to the project team at one of the workshop meetings. The participants found Mr. Wiemken's presentation to provide key insights into these important issues. With Mr. Wiemken's permission, his presentation to the project workshop and a later related white paper that he developed are included in the appendix. The white paper is reprinted from *Avoiding Rate Shock: Making the Case for Water Rates*, used by permission. Copyright © 2004, American Water Works Association.

REFERENCES

American City and County. Municipal Cost Index. Available: <http://images.americancityandcounty.com/files/115/mcihistory.html>.[cited December 9, 2002].

American Water Work Service Co., Inc. May 2002. "Deteriorating Buried Infrastructure Management Challenges and Strategies." Available: <http://www.epa.gov/ogwdw000/tcr/pdf/infrastructure.pdf>. [cited December 9, 2002].

American Water Works Association. 2001a. *Dawn of the Replacement Era—Reinvesting in Drinking Water Infrastructure.* Denver, CO.

American Water Works Association. 2001b. *Rehabilitation of Water Mains (M28).* Denver, CO.

American Water Works Association. *Principals of Water Rates, Fees, and Charges (AWWA M1).* Denver, CO.

Association of Metropolitan Sewerage Agencies and Association of Metropolitan Water Agencies. 1998. *Thinking, Getting, Staying Competitive: A Public Sector Handbook.* Washington, DC

Association of Metropolitan Sewerage Agencies. 2002. *Managing Public Infrastructure Assets To Minimize Cost and Maximize Performance.* Washington, DC.

Australian Infrastructure Report Card. 2001. Available: <http://www.InfrastructureReportCard.org.au>. [cited December 9, 2002]

Australian National Audit Office. 1996. *Asset Management Handbook.*

Banyard, J.K. 2000. "Asset Management: The Challenge for the 21st Century." AWWA Infrastructure Conference Proceedings.

Beaudet, B., T. Chavez, M. Matichich, R. McClintock, J. Rogers, A. Tokuda, and G. Wammock. 2001. Capital Planning Strategy Manual [CD-ROM]. Available: AwwaRF and AWWA. [cited December 9, 2002].

Board on Infrastructure and the Constructed Environment. 1995. *Measuring and Improving Infrastructure Performance.* Washington, DC: National Academic Press.

Brueck, Terrance M. 2000. *Developing and Implementing a Performance Measurement System: Volume I.* Alexandria, VA: Water Environment Research Foundation.

Burlingame, G., J. Rahman, E. Navera, and J.E. Durrant. 1998. Managing Water Supply Infrastructure. *Jour. AWWA*, 90(7):53.

Campbell, P., A. Judet, and D. Fortune. 2000. "Integrated Reporting Systems for Effective Asset Management." AWWA—Infrastructure Conference Proceedings.

Canning, M.E. 2001. "Field Testing of Water Distribution Infrastructure in the Asset Evaluation Process. AWWA—Infrastructure Conference Proceedings.

Causey, P.H., R.M. Musgraves, and M.S. Gotshall. October, 1997. "Comprehensive Capital Replacement Planning: Why Give It Another Thought?." WEF Management: Planning for Change.

Charlton, R., and M. Matichich. 2001. "Financing Substantial Capital Programs When Your Customer Base is Not Growing." AWWA/WEF—Joint Management Conference Proceedings.

Clemons, E.K. "Using Scenario Analysis to Manage the Strategic Risks of Reengineering." Sloan Management Review. Summer 1995.

Conroy, P.J., and D.M. Hughes. June 1997. "Cost Effective Water System Rehabilitation Through Whole Life Costing Analysis." AWWA—Annual Conference Proceedings.

Cromwell, J., G. Nestel, R. Albani, L. Paralez, A. Deb, and F. Grablutz. 2001. *Financial and Economic Optimization of Water Main Replacement Programs.* Denver, CO: AwwaRF.

Crotty, P. 2004. *Selection and Definition of Performance Indicators for Water and Wastewater Utilities.* Denver, CO: AwwaRF.

Deb, A.K., J.K. Snyder J.J. Chelius, J. Urie, and D.K. O'Day. 1990. *Assessment of Existing and Developing Water Main Rehabilitation Practices*. Denver, CO: AwwaRF and AWWA.

Deb, A.K., Y. Hasit, F. Grablutz, and R. Herz. 1998. *Quantifying Future Rehabilitation and Replacement Needs of Water Mains*. Denver, CO: AwwaRF and AWWA.

Deb, A.K., Y. Hasit, C. Norris. 1999. *Demonstration of Innovative Water Main Renewal Techniques*. Denver, CO: AwwaRF and AWWA.

Deb, A.K., F.M. Grablutz, Y.J. Hasit, J.K. Snyder, G.V. Loganathan, and N. Agbenowosi. 2002a. *Develop Decision Criteria to Prioritize Replacement and Rehabilitation of Mains and Appurtenances*. Denver, CO: AwwaRF and AWWA.

Deb, A.K., Y. J. Hasit, H. M. Schoser, J. K. Snyder, G.V. Loganathan and P. Khambhammettu 2002b. *Decision Support System for Distribution System Piping Renewal.* Denver, CO: AwwaRF.

deMansabert, S., and P. Thornton. "A Benders Decomposition Model for Sewer Rehabilitation Planning for Infiltration and Inflow Planning." Water Environment Research. March/April 1997.

Doyle, M.J., and D. Rose. July 2001. "Protecting Your Assets." *Water Environment & Technology*. Vol. 13, No. 7

Elliott, L., M. Stecklein, M. Martin, C. Feizoulof, M. Fisher, P. Giorsetto, J. Jacobs, B. O'Donnell, G. Sagehorn. 2001. *Water Treatment Plant Infrastructure Assessment Manager: User's Manual*. Denver, CO: AwwaRF.

Engineering News-Record. "Construction Cost Index History (1908-2002)" Available: <http://www.enr.com/cost/costcci.asp>

Gohier, L. "Why We Need an IAMS." APWA International Public Works Congress NRCC/CPWA Seminar Series "Innovations in Urban Infrastructure" Available: <http://www.nrc.ca/irc/fulltext/apwa/apwaiams.pdf>. [cited December 9, 2002]

Governmental Accounting Standards Board, Financial Accounting Foundation, No. 171-A, *Statement No. 34 of the Governmental Accounting Standards Board: Basic Financial Statements—and Management's Discussion and Analysis—for State and Local Governments*. June 1999.

Governmental Accounting Standards Board, Financial Accounting Foundation, Implementation Guide, Guide to Implementation of GASB Statement No. 34 on Basic Financial Statements—and Management's Discussion and Analysis—for State and Local Governments. 2000.

Governmental Accounting Standards Board, Financial Accounting Foundation, Implementation Guide, Guide to Implementation of GASB Statement No. 34 and Related Pronouncements: Questions and Answers. 2001.

Hajnosz, A. M., R. Cuthbert, and A. Hathiramani. 2001. "GASB 34: A Systematic Approach to Estimating Asset Replacement Costs." AWWA/WEF— Joint Management Conference Proceedings.

Harlow, V.K., and A.V. Czorny. October 2000. "Planning and Funding Asset Replacement: One Agency's Experience." WEFTEC 2000 Conference Proceedings.

Herbert, H. 1994. Technical and Economic Criteria Determining the Rehabilitation and/or Renewal of Drinking Water Pipelines. IWSA Regional Conference, Zurich, 16–20 May, 1994. 111–123.

Independent Pricing and Regulatory Tribunal of New South Wales. June 2001. "Water Industry Overview 2000." Available: <http://www.ipart.nsw.gov.au/pdf/op-9.pdf>. [cited December 9, 2002]

Johnstone, G.W. "Long-Term Planning for Capital Improvements." AWWA Proceedings AWWA Seminar on Demand Forecasting and Financial Risk Assessment. June 1985.

Jordan, J.L., C.N. Carlson, and J.R. Wilson. August 1997. Financial Indicators Measure Financial Health. *Jour. AWWA,* 89(8):34.

Kaplan, R.S., and D. P. Norton. 1996. "Using the Balanced Scorecard as a Strategic Management System." In Harvard Business Review, January-February.

Kleiner, Y., and B. Rajani. July 1997. Using Limited Data to Access Future Needs. *Jour. AWWA,* 91(7):47.

Landry, C.J. 2000. "New Economy of Water." Water Resources Impact. 4(1), pg 2-3. Available: <http://www.awra.org/impact/0201impact.pdf>. [cited December 9, 2002]

Loftin, K.M. and J. Rogers. "A New Approach in Risk Management and Capital Improvement Planning." AWWA 1997 Annual Conference Proceedings. June 1997.

Loganathan, G.V., H.D. Sherali, S. Park, and S. Subramanian, February 2001. *Optimal Design-Rehabilitation Strategies for Reliable Water Distribution Systems.* Final report submitted to the U.S. Geological Survey.

Matichich, M., and R. Schwarzwalder. 2000. "Asset Management for the New Millennium: Strategic Approaches for Water Utilities." AWWA—Annual Conference Proceedings.

McConnell, S., and Construx Software. 1998. *Software Project Survival Guide: How To Be Sure Your First Important Project Isn't Your Last.* Edited by D. Clark and V. Thulman. Redmond, Washington: Microsoft Press.

Morgan, T.R., E. Wagner, and R. Rowe. July 2001. "Asset Preservation and Risk Management for Collection Systems." 2001 A Collections Systems Odyssey: Integrating O&M and Wet Weather Solutions.

Morgan, T.R., and E. Wagner. July 2000. "Infrastructure Capital Assets Management." Watershed 2000 Management Conference.

Naumick, G., S. Chiavari, J. Rogers, and A. Tokuda. 2001. "A Systematic Approach to Prioritizing Water Utility Capital Spending." AWWA/WEF—Joint Management Conference Proceedings.

New Zealand Water and Wastes Association. 1999. "New Zealand Infrastructure Asset Grading Guidelines—Water Assets." Onehunga, Auckland: New Zealand Water and Wastes Association, Inc.

NZ National Asset Management Steering Group and the Institute of Public Works Engineering of Australia. April 2000. *International Infrastructure Management Manual—Australia/ New Zealand Edition.* Johnsonville, Wellington: Association of Local Government Engineers New Zealand.

O'Day, D.K., R. Weiss, S. Chiavari, and D. Blair. 1986. Water Main Evaluation for Rehabilitation/Replacement. Denver, CO: AWWARF and AWWA.

Paper, E. SuperBuild: Building Ontario's Future. Available: <http://www.gov.on.ca/FIN/ bud01e/pdf/papere.pdf>. [cited December 9, 2002]

Paralez, L.L., and D. Muto. 2001. "Creating an Asset Management Strategy: An Asset Management Template." AWWA— Infrastructure Conference Proceedings.

Patton, T.K., and P.S. Wardlow. May 1999. "Why Infrastructure Reporting?" GASB Action Report Vol. 16, No. 5. Available: <http://www.gasb.org/repmodel/index.html>. [cited December 9, 2002]

Pickering, D., and S. Leek. "Integrating Risk Analysis With Asset Management Decision Making." AWWA—Annual Conference Proceedings.

Pullen, S.F. 1999. "Consideration of Environmental Issues When Renewing Facilities and Infrastructure." 8th International Conference on Durability of Building Materials and Components May 30-June 3, 1999. Available: <http://auslcanet.rmit.edu.au/Papers /8dbmcpaper.doc>. [cited December 9, 2002].

Purvenas, T.J. 1998. Infrastructure Replacement—Credit Quality Concerns, in *Water*. Washington, DC: National Association of Water Companies.

Rothstein, E.P., and D.G. Burna. 1997. "Prioritization of Water Utility Capital Spending: An Analysis Framework." AWWA—Annual Conference Proceedings.

Sivalingam, Y. 2001. "GASB 34 Regulations and Asset Infrastructure Management." AWWA/WEF—Joint Management Conference Proceedings.

Spacek, J., and M. Matichich. 2001. "How to Gain Funding Support for Distribution System Renewal and Replacement Programs." AWWA—Annual Conference Proceedings.

Speranza, E., M. Matichich, G. Wammock, V. Ragucci, D. Green, J. Silverberg, and R. Rosholt. 2004. Avoiding Rate Shock: Making the Case for Water Rates. American Water Works Association.

Standards Australia Code, AS/NZS 4360:1999 Risk Management

Standish Group. 1994. Charting the Seas of Information Technology. Dennis, Mass.

Styles, M.D.G. 2000. "Best Value Advanced Asset Management—An Australian Perspective." APWA International Public Works Congress NRCC/CPWA Seminar Series "Innovations in Urban Infrastructure" Available: <http://www.nrc.ca/irc/fulltext/apwa/apwa2000 styles.pdf>. [cited December 9, 2002]

Thornton, B., and J. Rogers. September/October 1999. "Management of Capital Infrastructure Assets Keeps Utilities Ahead of the Game" *Utility Executive.*

Thornton, W.J., G. Compton, D. Song, and M. Brandon. 2001. "Risk Assessment Process Aids in Prioritizing Water Distribution System Remediation." AWWA/WEF—Joint Management Conference Proceedings.

U.S. Air Force Civil Engineers, *Operational Risk Management Handbook*

U.S. Air Force Instruction 91-213, *Operational Risk Management Program*

U.S. Air Force Pamphlet 91-214, *Operational Risk Management Implementation and Execution*

U.S. Air Force Pamphlet 91-215, *Operational Risk Management Techniques and Tools*

USEPA. 1997. Drinking Water Infrastructure Needs Survey: First Report to Congress. EPA-812-R-97-001. Washington, DC: USEPA.

USEPA. 1998. "Urban Infrastructure Research Plan: Water and Wastewater Issues." (Washington, DC: Urban Watershed Management Branch). Available: <http://www.epa.gov/ednnrmrl/repository/Infrastructure_plan/urban_infra_res_plan.pdf>. [cited December 9, 2002]

USEPA. April 2000. "Conservation Pricing of Water and Wastewater." Available: <http://www.epa.gov/owm/cwfinance/cwsrf/consrvprice.pdf>. [cited December 9, 2002]

USEPA. August 2000. “The Drinking Water State Revolving Fund Program Case Studies in Implementation III. Disadvantaged Communities.” EPA 816-R-00-005. (Washington, DC: Office of Groundwater and Drinking Water).

USEPA. February 1998. “Information for States on Developing Affordability Criteria for Drinking Water.” EPA 816-R-98-002. (Washington DC: Office of Water).

USEPA. September 2002. “The Clean Water and Drinking Water Infrastructure Gap Analysis.” EPA-816-R-02-020. (Washington, DC: Office of Water). Available: <http://www.epa .gov/OGWDW/gapreport.pdf>. [cited December 9, 2002].

US GAO. August 2002. Water Infrastructure Information on Financing, Capital Planning, and Privatization. GAO-02-764. Washington, DC: Natural Resources and Environment.

US GAO. September 1999. Military Infrastructure Real Property Management Needs Improvement. NSIAD-99-100. Washington, DC: National Security and International Affairs Division.

Walski, T.M. 1985. Cost of Water Distribution System Infrastructure Rehabilitation, Repair, and Replacement. Technical Report EL-85-5. Vicksburg, MS: U.S. Army Engineer Waterways Experiment Station.

Water Environment Federation. 1999. *Wastewater Collection Systems Management, MOP 7, 5th Edition.*

ABBREVIATIONS

AAA	American Appraisal Associates
AMU	asset management unit
AWWA	American Water Works Association
AwwaRF	Awwa Research Foundation
CD-ROM	compact disc—read-only memory
CIP	capital improvement plan
CMMS	computerized maintenance management system
EPA	Environmental Protection Agency
GAO	General Accounting Office
GASB	Government Accounting Standards Board
GIS	geographic information system
IIMS	Integrated Infrastructure Management System
MMS	maintenance management system
O&M	operation and maintenance
OCL	sodium hypochlorite
PI	principal investigator
PWTP	Penitencia Water Treatment Plant
R&R	renewal and rehabilitation
RWTP	Rinconada Water Treatment Plant
S&P	Standard & Poor
SCVWD	Santa Clara Valley Water District
SEWL	South East Water LLC
SFPUC	San Francisco Public Utilities Commission
SPU	Seattle Public Utilities
SRF	State Revolving Fund
STWTP	Santa Teresa Water Treatment Plant
TIADS	Toronto Infrastructure Asset Data Standard
WTP	water treatment plant

IWA Publishing's authorised EU representative for General Product Safety Regulations is Diane D'Arras, 15 rue Duret, 75116 Paris, France, e-mail: safety@iwap.co.uk.

Printed and bound by CPI Group (UK) Ltd, Croydon, CR0 4YY
17/09/2026
02218802-0016